T0257847

Manufacturing System Advances

Edited by **Jeff Hansen**

CLANRYE
INTERNATIONAL

New Jersey

Published by Clanrye International,
55 Van Reypen Street,
Jersey City, NJ 07306, USA
www.clanryeinternational.com

Manufacturing System Advances
Edited by Jeff Hansen

© 2015 Clanrye International

International Standard Book Number: 978-1-63240-334-6 (Hardback)

Printed in the United States of America.

Contents

Preface

This book discusses current advances, devices, methods and novel ideas in manufacturing systems. It consists of equipment, products, people, knowledge, control and support functions so as to create a framework which caters to market needs and demands in today's competitive world. It collects various researches conducted in fundamental and applied industrial fields. The book covers the latest developments and techniques and systems such as approach based on response surfaces method and ant colony system, wafer fabrication, fundamentals of global outsourcing for manufacturers, and hybrid manufacturing system design and development. The book will be useful to those involved in manufacturing engineering, systems and management and in manufacturing research.

All of the data presented henceforth, was collaborated in the wake of recent advancements in the field. The aim of this book is to present the diversified developments from across the globe in a comprehensible manner. The opinions expressed in each chapter belong solely to the contributing authors. Their interpretations of the topics are the integral part of this book, which I have carefully compiled for a better understanding of the readers.

At the end, I would like to thank all those who dedicated their time and efforts for the successful completion of this book. I also wish to convey my gratitude towards my friends and family who supported me at every step.

Editor

An Approach-Based on Response Surfaces Method and Ant Colony System for Multi-Objective Optimization: A Case Study

Barthélemy H. Ateme-Nguema[1], Félicia Etsinda-Mpiga[2],
Thiên-My Dao[2] and Victor Songmene[2]
[1]*Université du Québec en Abitibi-Témiscamingue*
[2]*École de Technologie Supérieure, Montreal,*
Canada

1. Introduction

Numerical optimization is the search approach we adopted in order to determine the best mechanical design. Several algorithms related to the problems at hand were developed, most of them single-objective. However, given the complexity of the products involved and the multiple objectives of the design considered, the researchers focused on the optimization algorithms for such problems. In short, the optimization problems have multiple objectives, and in many cases, there are multiple constraints. Design processes often require expensive evaluations of objective functions. That is particularly the case when such performance indexes and their constraints are obtained through intermediate simulations by finite elements involving fine meshes, many freedom degrees and nonlinear geometrical behaviours. To overcome these difficulties, the response surface method (RSM) is employed (Myers & Montgomery, 2002; Roux et al., 1998; Stander, 2001; Zhang et al., 2002) to replace a complex model by an approximation based on results calculated on certain points in search space.

When an adequate model is obtained with the RSM approach, it then becomes necessary to consider the optimization step. The method used to find the best solution assesses several objectives simultaneously; since some such objectives are fundamentally conflicting vis-à-vis of another, we therefore need to establish a compromise. Existing literature shows that desirability or metaheuristic functions are normally used, the most common being the genetic algorithm (GA). Sun & Lee (Sun & Lee, 2005), present an approach which associates the RSM and GA with the optimal aerodynamic design of a helicopter rotor blade. The ACO is a metaheuristic, which has been successfully used to solve several combinatorial optimization problems. We however see that very little exists in terms of documentation for optimization using ACO, as far as multiobjective problems are concerned. Some works lead us to believe that ant colonies can produce an optimum situation faster than the GA (Nagesh, 2006; Liang, 2004). In the literature, ACOs are used almost exclusively for *"Travelling Salesman Problem"* (TSP), quadratic assignment problem allocation (QAS), constraint satisfaction problems (CSP), design manufacturing systems (DMS), and for discrete and combinatorial optimization problems. Our contribution consists in an extension of the ACO in the multiobjective optimization of mechanical system design in a continuous field. This paper starts with the modelling process with RSM, and then goes on to describe the ACO and the Hybrid method developed for a problem regarding multiobjective optimization with constraints. An application of the suggested method for optimizing the mechanical process design is presented.

2. Modeling with RSM

RSM is a collection of statistical and mathematical techniques used to develop, improve and optimize processes (Myers & Montgomery, 2002). Furthermore, it has important applications in the design and formulation of new products, as well as in the improvement of existing products.

The objective of RSM is to evaluate a response, i.e., the objective physical quantities, which are influenced by several design variables. When we use RSM, we seek to connect a continuous answer Y with continuous and controlled factors X_1, X_2... X_k, using a linear regression model which can be written (Myers & Montgomery, 2002) as:

$$y = f_\beta(X_1, X_2, ..., X_p) + \varepsilon \tag{1}$$

Since the response surface is described by a polynomial representation, it is possible to reduce the optimization resolution process time by assessing the objectives with their models rather than using more complex empirical models such as those obtained through the FEM analysis. Although the specific form of response factor f_β is unknown, experience shows that it can be significantly approximated using a polynomial.

In the case of two factors, the linear regression model is one of the simplest available, and corresponds to a first-degree model with interaction, and which has the following form:

$$y = \beta_0 + \beta_1 X_1 + \beta_2 X_2 + \beta_{12} X_1 X_2 + \varepsilon \tag{2}$$

Whenever this model is unable to describe the experimental reality effectively, it is common practice to use a second-degree model, which includes the quadratic effects of the factors involved:

$$y = \beta_0 + \beta_1 X_1 + \beta_2 X_2 + \beta_{12} X_1 X_2 + \beta_{11} X_1^2 + \beta_{22} X_2^2 + \varepsilon \tag{3}$$

Where y is the response (study objective, for example, the total manufacturing cost); ε is the estimate of the error; X_1 and X_2 are influential factors of the coded response (e.g., design variables).

The unknown parameters of this mathematical model, βi values, are estimated through the least-squares technique, and the adjustment quality of the model is assessed using traditional multiple linear regression tools.

Ideally, the number of experiments carried out, either with the finite element model (FEM) or using other approaches, during the application of RSM, should be as small as possible, in order to reduce data-processing requirements. Properly selecting the points to be used for the simulation will allow a reduction of the variance of the coefficients of the mathematical model, which will in turn ensure that the response surfaces obtained are more reliable. To that end, we need to determine the experimental design to be adopted in order to obtain the most interesting simulation for this problem. The central composite design (CCD) was employed in the case of the second-order response surface, but other types of plans, such as the complete factorial design and the fractional factorial design, are also available for use.

Once the mathematical models are obtained, we need to verify that they produce an adequate approximation of the actual study system. The statistic selection criterion is the coefficient of determination R^2, which must be as close as possible to 1 ($0 < R^2 < 1$).

Once this stage is completed, we will have all the equations which make up our multiobjective optimization problem. Generally, such problems are as the following form:

$$
\begin{aligned}
&\text{Find} && x = \left[x_1, x_2, x_3, \ldots, x_n\right]^T \\
&\text{Which minimize} && f(x) = \left\{f_1(x), f_2(x), f_3(x), \ldots, f_n(x)\right\} \\
&\text{Subject to} && g_j(x) \leq 0 \qquad \text{for } j = 1, m \\
& && x_i^L \leq x_i \leq x_i^U \qquad \text{for } i = 1, n
\end{aligned}
\tag{4}
$$

To optimize this problem, we explored the ant colony algorithm (ACO). Some options are offered for this kind of problem, such as the desirability function and the genetic GA used by some authors, such as Sun & Lee (Sun & Lee, 2005) or Abdul-Wahad & Abdo (Abdul-Wahad & Abdo, 2007). The literature shows that for many problems, the ant colony approach produces better results in terms of quality solutions and resolution speed, as compared to the GA. This allowed us to begin this research with the resolution of the multiobjective continuous optimization problem in mechanical design.

3. Ant colony algorithm approach

The ACO metaheuristic, called the ant system (Dorigo, 1992), was inspired by studies of the behaviour of ants (Deneubourg et al., 1983; Deneubourg & Goss, 1989; Goss et al., 1990), as a multi-agent approach for resolving combinative optimization problems such as the TSP.

Ants communicate among themselves through the "*pheromone*", a substance they deposit on the ground in variable amounts as they move about. It has been observed that the more ants use a particular path, the more pheromone is deposited on that path, and the more attractive it becomes to other ants seeking food. If an obstacle is suddenly placed on an established path leading to a food source, ants will initially go randomly right or left, but those choosing the side that is in fact shorter will reach the food more quickly, and will make the return journey more often. The pheromone concentration on the shorter path will therefore be more strongly reinforced, and it will eventually become the new preferred route for the stream of ants; however, it must also be borne in mind that the pheromone deposited along the way does evaporate. Works by Colorni et al. (Colorni et al., 1992), Dorigo et al. (Dorigo et al., 1996; Dorigo et al., 1999; Dorigo et al., 2000) provide detailed information on the operation of the algorithm and on the determination of the values of the various parameters (see Fig. 1).

In our field, the ACO has been used very sparingly, and has been focused primarily on single-objective problems (Chegury, 2006). For multiobjective problems, the ACO has hardly been used at all (Zhao, 2007), and when used, has been mainly on combinatorial optimization problems. The importance of this work therefore lies in its attempt to adapt continuous ant colonies to multiobjective problems.

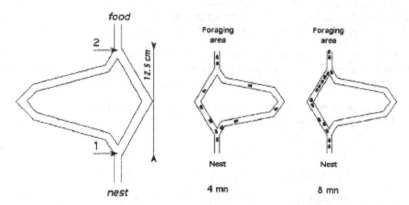

Fig. 1. Experimental setup and drawings of the selection of short branches by a colony of "*Linephitema humile*", 4 and 8 min after the bridge was placed (Dorigo et al., 2000)

4. Proposed design optimization approach

The objective of this chapter is to determine the best design for a mechanical system such as a plane wing, an engine, etc., or for an unspecified mechanical process that sometimes simultaneously optimizes several conflicting objectives. The ACO, like the GA, requires an objective function which can be quickly assessed. We use RSM modeling to determine such objective functions, and the ACO as the research method. Reducing the resolution time in the optimization process requires a reduction of the preciseness of the assessment of objective functions, since we use an approximate modeling of our objectives instead of their exact representations.

Each objective f_i is expressed according to the variables of real design x_i which influence its value. The multiobjective optimization model obtained with RSM is:

$$\text{Minimize} \quad f(x) = \{f_1(x), f_2(x), f_3(x), ..., f_n(x)\}$$
$$\text{subject to} \quad x_i^L \le x_i \le x_i^U \quad \text{for } i = 1, n \tag{5}$$

After obtaining a mathematical model for that problem, the optimization phase must be able to determine the best compromise solution for the various objectives.

The steps for the general ACO metaheuristic for compromise solutions for combinatorial problems presented by Gagné et al. (Gagné et al., 2004) constitute an interesting approach to be considered in our resolution process for developing the fitness function.

4.1 Continuous ant colonies

There are several ant colony algorithms available for continuous optimization, the first of which was developed by Bilchev & Parmee (Bilchev & Parmee, 1995), and named CACO

(Continuous Ant Colony Optimization), using ant colonies for local searches and calling upon revolutionary algorithms for global searches. Ling et al. (Ling et al., 2002) present an unspecified hybrid algorithm whose main premise is to consider the differences between two individuals on each dimension as many parts of a path on which the pheromones are deposited. The evolution of the individuals dealt with mutation and crossing-over operators. This method thus tries to reproduce the construction mechanism of the solution component by component.

Monmarché et al. (Monmarché et al., 2000) developed the API algorithm which takes the primitive ant behaviour of the species *Pachycondyla Apicalis*, and which does not use indirect communication by tracks of pheromone. In this method, it is necessary to start by positioning a nest randomly on the research space, after which ants are distributed randomly over it. These ants explore their *"hunting site"* locally by evaluating several points within a given perimeter. Socha (Socha, 2004) presents the ACO algorithm for continuous optimization which tries to maintain the iterative construction solutions for continuous variables. He considers that the components of all solutions are formed by the various optimized variables. Moreover, before considering the algorithm from the ant's point of view, he opts to operate at the colony level, with the ants being simply points to be evaluated. Pourtakdoust & Nobahari (Pourtakdoust & Nobahari, 2004) developed the CACS (Continuous Ant Colony System) algorithm, which is very similar to that of Socha. Indeed, in CACS, as is the case with ACO, for continuous optimization, the core of the algorithm consists in evolving a probability distribution which for CACS is normal.

4.2 Proposed algorithm

Once the steps used in making a choice regarding the elements to be included in the resolution process are explained. We present the new proposed algorithm for our approach (see Fig. 2a and Fig. 2b).

Step 1: System configuration

Determine the objectives of this study, the constraints and the variables which can influence these objectives. Evaluate the field of application of these variables.

Step 2: RSM

Set up an experimental design, carry out tests, and model the various objectives according to influential parameters.

Step 3: Seek ideal point

Using RSM, determine distinct optimum for each study objective.

Step 4: Optimization function formulation

a. State user preferences (weighting of the objectives).
b. The various objectives are expressed in a single function: the fitness function. It acts as an equation which for each objective, expresses the standard and balanced distance at the ideal point F^* of an unspecified solution k, whose various objectives are given by F^k. This function makes it possible to standardize objectives in order to reduce the adverse effects obtained from the various measuring units, as well as the extent of the field of the variables, in order to not skew the fitness function (Gagné et al., 2004):

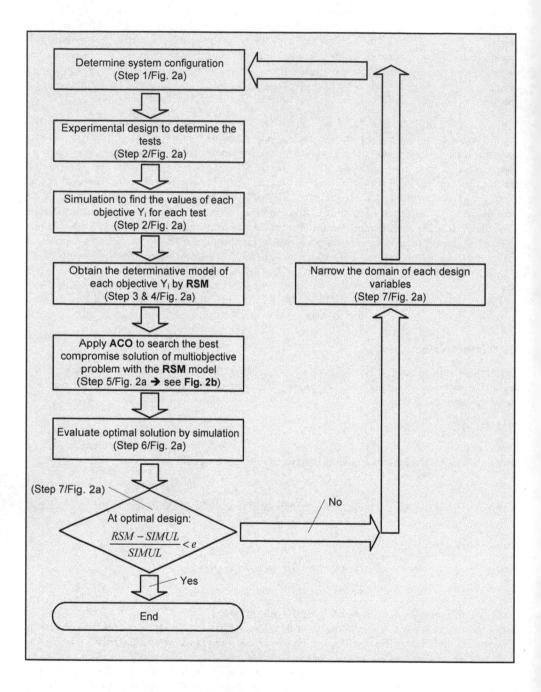

Figure 2a. Flow chart of the optimization approach

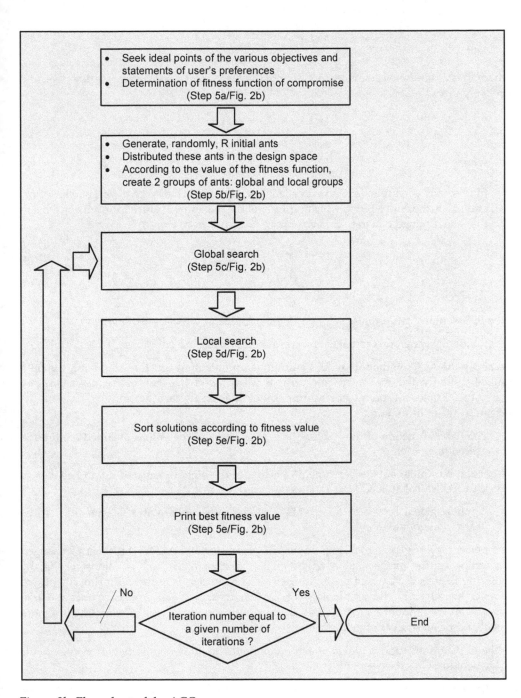

Figure 2b. Flow chart of the ACO process

$$fitness = \left(\sum_{i=1}^{Z} p_i \left(\frac{F_i^k - F_i^*}{F_i^{nad} - F_i^*} \right)^2 \right)^{\frac{1}{2}} \quad (6)$$

where F^* is a solution vector corresponding to the ideal point of each separate objective, and probably expressed by:

$$F^* = \left\{ F_1^*, F_2^*, ..., F_Z^* \right\} \quad (7)$$

where $F_i^* = \min_{x \in S} f_i(x)$.

F^* generally corresponds to an unrealizable solution. S is the space of acceptable search, and F^{nad} is the Nadir point, which represents the maximum values for each objective in the set of optimal Pareto solutions:

$$F^{nad} = \left\{ F_1^{nad}, F_2^{nad}, ..., F_z^{nad} \right\} \quad (8)$$

where $F_i^{nad} = \max_{x \in S^*} f_i(x)$.

Step 5: Determine compromise solution

a.   Randomly generate R initial ants corresponding to feasible solutions.

To apply the ACO methodology for continuous optimization function problems, the field must be subdivided into a specific area, R, distributed by chance. Next, we need to generate feasible solutions representing the initial ants, each forming a part of the research area to be explored.

b.   The fitness function of these solutions is assessed, and the values obtained are lines in descending order.

We obtain our initial ants "R" and the proportion of the higher values of R will be taken to constitute the global ants "G".

c.   Apply a global search to a percentage of the initial ants, with "G" constituting the "*worst*" solutions available.

The percentage of global ants is an important parameter of CACO, which can be changed depending on the problem at hand. A global search creates new solutions for "G" by replacing the weaker parts of the existing field. This process is composed primarily of two genetic operators. In the terminology of CACO, these are called *random walk and trail diffusion*. In the random search process, the ants move in new directions in search of more recent and richer sources of food.

In the CACO simulation, a global search is conducted in all fields through a process that is equivalent to a GA crossover and mutation.

-    *Crossover or random walk:* The crossover operation is conducted to replace inferior solutions with superior ones, with the crossover probability (CP).

- *Mutation*: The replaced solutions are further improved by mutation. The mutation step is completed in CACO by making an addition or proportional subtraction to the mutation probability. The mutation step size is reduced or increased as per Eq. 9. (Mathur et al., 2000)

$$\Delta = R(1 - r^{(1-T)^b})$$
(9)

where r is a random number from 0 to 1, R is the maximum step size, T is the ratio of the current iteration number and that of the total number of iterations, and b is a positive parameter controlling the degree of nonlinearity.

- *Trail diffusion*: In this step, the field of the global search is gradually reduced, as the search progresses. This reduction makes it possible to increase the probability of locating the optimum through more concentric search procedures. Trail diffusion is similar to the arithmetic GA crossover. In this step, two parents are randomly selected from the parent population space. The elements of the child's vector can be any one of the following:
1. The child corresponds to an element from the first parent
2. The child corresponds to an element from the second parent
3. The child is a combination of the parents (Eq.10) (Mathur et al., 2000)

$$X(child) = (\alpha) X_{i(parent1)} + (1 - \alpha) X_{i(parent2)}$$
(10)

where α is a uniform random number ranging from [0 to 1]

The probability of selecting one of the three options depends on the mutation probability. Thus, if the mutation probability is 0.5, option 3 can be selected with a probability of 50%, whereas the probability of selecting option 1 or 2 is 25%.

d. Send local ants L in the various R areas

Once the global search is completed, the zones to which you send the local ants are defined and a local search can begin.

In a local search, the local ants choose the area to be explored among the areas of the matrix R, according to the current quantity of pheromones in the areas. The probability of choosing a solution "i" is given by: (Mathur et al., 2000)

$$P_i(t) = \tau_i(i) \Big/ \sum_k \tau_k(t)$$
(11)

where "i" is the solution index and $\tau_i(t)$ is the pheromone trail on the solution "i" at time "t".

After choosing its destination, the ant proceeds across a short distance. The search direction remains the same from one local solution to the next as long as there is improvement in the fitness function. If there is no improvement, the ant reorients itself randomly to another direction. If an improvement in the fitness function is obtained in the preceding procedure, the position vector of the area is updated. The quantity of pheromone deposited is proportional to the improvement of the fitness function. If, in the search process, a higher fitness function value is obtained, the age of the area is increased. This age of the area is another major parameter in the CACO algorithm. The size of the ant displacement in a local search depends on the current age. The search ray is maximum for age zero, and minimal for the maximum age, with a linear variation.

e.   Evaluate the fitness function for each ant obtained, and continue the iterative process, beginning with a global search until stop conditions are observed.

Step 6: Evaluate the best solutions (quasi optimal) by simulation or experimentation for the experimental design (Example, FEM)

Step 7: Evaluate the stop criterion $\dfrac{RSM - SIMUL}{SIMUL} < e$ with SIMUL being the simulation result.

In the optimization design problem for a mechanical system, the number of design variables is very often equal to or higher than 3, and each one of them has a broad field of variation. Consequently, in our resolution process, it is possible for the search field for each design variable to be gradually narrowed for as long as the stop criterion has not been encountered. The search process ends when $\dfrac{RSM - SIMUL}{SIMUL} < e$ with e being a margin of error defined beforehand.

5. Application: Numerical example of two-objective problem

In order to illustrate the performances of the recommended resolution approach used in this paper, we carried out the optimization of a multistage flash desalination process. The problem was taken from Abdul-Wahab & Abdo (Abdul-Wahab & Abdo, 2007), and was solved using the experimental designs, and optimized using desirability functions.

5.1 Problem definition

Multistage flash (MSF) desalination is an evaporation and condensation process, which involves boiling seawater and condensing the vapour to produce distilled water. A more extensive description of the multistage flash desalination MSF considered in this work can be found in Hamed et al. (Hamed et al., 2001).

In this study, two performance objectives are considered: the maximization of the distillate produced rate (DF) and the minimization of the blow down flow rate (BDF). The operation variables which influence these objectives are presented in Table 1 (**Step 1**). They include:

Parameter name	Nomenclature		Low level	High level
Seawater inlet temperature (°C)	SWIT	(A)	24	35
Temperature difference (°C)	TD	(B)	5.2	8.0
Last-stage brine level (mm)	LSBL	(C)	50	850
First-stage brine level (mm)	FSBL	(D)	40	320
Brine recycle pump flow (m3/h)	BRPF	(E)	8200	11 500

Table 1. Design parameters

5.2 Modeling with RSM

To express our objectives according to decision variables, we need to use modeling with RSM (**steps 2 and step 3**). We considered the experiments carried out by Abdul-Wahab & Abdo (Abdul-Wahab & Abdo, 2007), which helped us to design our model.

Abdul-Wahab & Abdo resorted to a two-level factorial design, carried out 64 experiments and five central-point tests with design variables coded on two levels: low (-1) and high (+1). The experimental design provides us with a linear regression model coded for each response in this study (see Fig. 3 & Fig. 4).

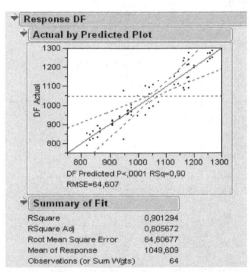

Fig. 3. Modeling with RSM – part I

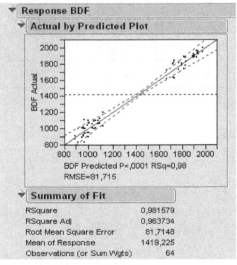

Fig. 4. Modeling with RSM – part II

Finally, the equations representing our objectives are:

$$
\begin{cases}
DF = 1041.61 + 20.45(B) + 18.65(C) + 120.29(E) \\
\qquad + 26.46(AC) + 30.08(CD) - 25.06(ABE) \\
BDF = \qquad 1419.22 + 414.54(C) + 34.77(D) \\
\qquad + 28.71(AC) - 25.63(ABD)
\end{cases}
\tag{12}
$$

We also note that in our experimental design, the variables C and E are the most influential on our objectives (see Fig. 5).

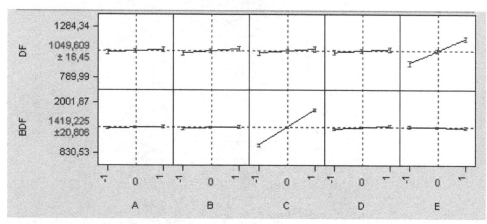

Fig. 5. Impact of the decision variables on study objectives

Additional information on optimization, as well as the goals of the study, is summarized in the following table:

Objectives	Goal	Lower limit	Upper limit	Weighting
BDF	To minimize	830.53	2001.87	3
DF	To maximize	789.99	1284.34	5

Table 2. Constraints on objectives of study

5.3 Multiobjective optimization

Let us optimize the following problem with our CACO multiobjective approach:

$$
\begin{aligned}
\text{Find} \quad & x = [A, B, C, D, E]^T \\
\text{which minimize} \quad & f(x) = \{-DF(x), BDF(x)\} \\
\text{subject to} \quad & DF(x) \leq 1284.34 \\
& DF(x) \geq 789.99 \\
& BDF(x) \leq 2001.87
\end{aligned}
\tag{13}
$$

The fitness function used, obtained by the CP (compromise programming) (Gagné et al., 2004) method, allows the search for solutions approaching the ideal point for each objective (**Step 4**):

$$fitness = \left(\frac{5}{8} * \left(\frac{DF_{max} - DF_i}{DF_{max} - DF_i^{Nad}} \right)^2 + \frac{3}{8} * \left(\frac{BDF_i - BDF_{min}}{BDF_i^{Nad} - BDF_{min}} \right)^2 \right)^{\frac{1}{2}} \tag{14}$$

The minimization of the fitness function enables us to reach our "BDF" minimization and "DF" maximization goals.

The result is a set of optimal Pareto solutions. We present more than one solution to the user in order to provide him with a margin of makeover. Abdul-Wahab & Abdo (Abdul-Wahab & Abdo, 2007), in their paper, present their 10 best solutions. We will do the same in order to make some comparisons.

Using the MatLab software (**step 5**), Figure 6 allows us to say that we get the best solution after 310 iterations. The staircase shape of the curve (Fig. 6) is explained by the memory effect that we used in the program code. Thus, when iteration produces a worse solution than the last one, this last solution (previous iteration) is retained.

The Table 3 presents the results obtained by the optimization process. The best solution is the number 1, while the 9 other solutions offer alternatives to user. These solutions meet problem constraints and, gives results which minimize "BDF" and maximize "DF" while remaining in the field of each decisional variable.

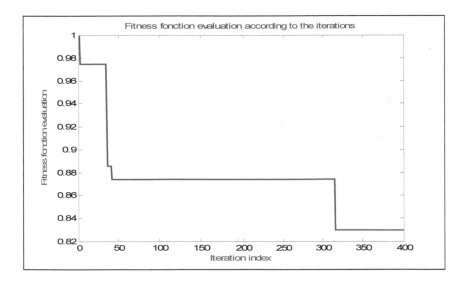

Fig. 6. Values of the fitness function according to the iteration count

# solutions	SWIT	TD	LSBL	FSBL	BRPF	DF	BDF	FITNESS
1	24.05	7.96	53.38	46.97	11499	1249.698	979.903	0.8295
2	24.26	7.99	66.23	44.787	11472	1245.603	990.740	0.8604
3	24.26	7.997	62.69	46.9	11492	1247.492	988.099	0.8738
4	24.37	7.94	54.20	48.49	11497	1245.577	981.335	0.8741
5	24.26	7.99	66.23	44.79	11472	1245.603	990.740	0.8854
6	24.61	7.99	53.99	43.67	11443	1245.552	985.796	0.8863
7	24.56	7.96	53.38	46.97	11499	1244.994	979.456	0.8904
8	24.33	7.94	50.86	71.3	11492	1240.785	987.596	0.8909
9	24.17	7.76	70.76	40.45	11439	1236.892	997.412	0.8963
10	24.72	7.97	60.13	40.12	11436	1239.208	982.990	0.8986

Table 3. Optimal solutions

The above Table (see Table 3) present the 10 best results of our study. These solutions meet the constraints of the problem and give excellent results which minimize "BDF" and maximize "DF" while remaining within the confines of each decision variable. It's interesting to observe the values of the decision variables in their respective fields. We can see that these best solutions are obtained under the following conditions (see Fig.7):

- Low temperature for seawater (SWIT) entering into the system
- The temperature difference (TD), which is high and similar for each of the solutions
- A final level of low salinity (LSBL)
- A first level of low salinity (FSBL)
- A high flow rate of the pump recycling salt (BRPF)

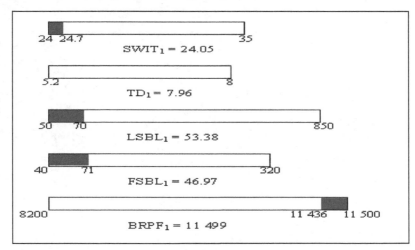

Fig. 7. Value margins of variables for optimal solutions

5.4 Comparison with authors' results

A comparison between the results obtained with the desirability function ("DF") and the hybrid approach developed ("$DF/CACO$ multiobjective") shows that the second gives better quality results. Recall that this comparison is made between the results obtained by the proposed approach and those of Abdul-Wahab & Abdo.

# of solution	BDF desirability	BDF-CACO multiobjective	% BDF improvement	DF desirability	DF CACO multiobjective	% DF improvement
1	991.725	979.903	1.19%	1224.52	1249.698	2.01%
2	1038.83	990.740	4.63%	1222.91	1245.603	1.82%
3	1033.34	988.099	4.386%	1213.11	1247.492	2.76%
4	1035.78	981.335	5.26%	1210.98	1245.577	2.78%
5	975.718	990.740	-1.54%	1181.75	1245.603	5.13%
6	965.137	985.796	-2.14%	1173.29	1245.552	5.80%
7	996.446	979.456	1.71%	1176.02	1244.994	5.54%
8	1035.8	987.596	4.65%	1166.87	1240.785	5.96%
9	973.657	997.412	-2.44%	1151.36	1236.892	6.92%
10	1005.05	982.990	2.19%	1141.94	1239.208	7.85%

Table 4. Results of multiobjective CACO versus the desirability function

Firstly, by observing the change in the response values we obtain for the various solutions (see Table 4), we can see that the solutions achieved with the hybrid approach vary much less than those obtained with the desirability function of Abdul-Wahab & Abdo (Abdul-Wahab & Abdo, 2007). It seems that our solutions are closer to each other. The reason is that the hybrid approach causes small displacements during the ant's research process. Thus when the fitness function decreases, the ants move over a short distance before re-test the function, if and only if, the obtained value is better than the previous one. Otherwise, the process reorients itself in case of declining performance.

Secondly and always in Table 4, by comparing our results with those of Abdul-Wahab & Abdo (Abdul-Wahab & Abdo, 2007) for the desirability function, the CACO-multiobjective approach shows that the second objective gives better quality results, with a 4.66% average improvement for the main goal (BDF & CACO-multiobjective), and 1.79% for the secondary one (DF & CACO-multiobjective).

Moreover, the observed variations in the answers values of the various solutions are visualized on figures 8a and 8b. (see below). These variations from the point of view of the BDF desirability function are shown on Fig 8a while those related to DF function are illustrated on Fig 8b. By observing these Figures, we observe that the solutions obtained with the CACO-multiobjective approach are smaller than those of classical approaches. As previously stated, these variations are explained by a small displacement of local ants, and when the fitness function decreases, ants move on a short distance before re-test the fitness function to obtain a new solution. These mechanisms and process and mechanisms are the same for the second desirability function visualized on Fig. 8b.

Following this application, and having obtained appreciable results, we can conclude that our algorithm functions correctly, while leading to coherent solutions, and that it has proven its effectiveness by obtaining better solutions than those of the authors, Abdul-Wahab & Abdo (Abdul-Wahab & Abdo, 2007).

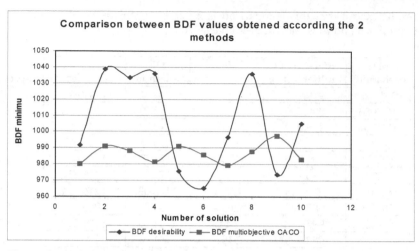

Fig. 8a. Chart of BDF values for optimal solutions (Part I)

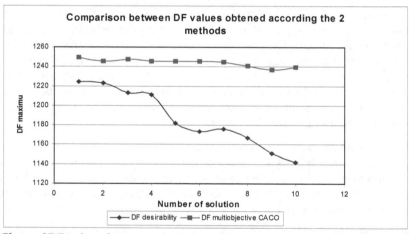

Fig. 8b. Chart of DF values for optimal solutions (Part II)

6. Conclusion

This book chapter presents a new multiobjective optimization approach for mechanical system design. Various techniques have traditionally been employed to resolve this kind of problem, including an approach combining RSM, GA and a simulation tool such as FEM. We have the ACO, which allows the exploration of a combination which includes another optimization algorithm. The ACO captured our interest because we were able to note in various works that in multiobjective optimization, it does produce better results than the quadratic programming technique and the GA. The ACO thus appears to be an innovative and leading solution for design optimization, because it is completely generalized and independent of problem type, which allows it to be modified in order to optimize the design of a complex mechanical system, subject to various economical and mechanical criteria, and respecting many constraints. However, it must be recalled that the ACO was developed to resolve discrete problems, and that its use on continuous problems is constantly under development; our study contributes to the development of the continuous ACO for multiobjective problems.

The approach we present makes it possible to effectively optimize a mechanical design problem. The approach performs much better when compared to using the desirability function. The results of the application allow it to validate the suggested design optimization method.

7. Acknowledgment

The author wishes to thank the co-authors of this chapter. Specifically, Miss Estinda-Mpiga for his major contribution and the quality of work. Thanks to all colleagues who have contributed from time to time, partially or more to obtain this manuscript.

Thank you to the Mechanical Engineering Department members at the *"École de Technologie Supérieure"* in Montreal and those of the Management Sciences Department at the University of Quebec in Abitibi-Témiscamingue (UQAT), Canada.

8. References

Abdul-Wahab, S.A. & Abdo, J. (2007). Optimization of multistage flash desalination process by using a two-level factorial design, *Applied Thermal Engineering*, Vol. 27, pp. 413–421

Bilchey, G. & Parmee, I.C. (1995). Ant colony metaphor for searching continuous design spaces, in: *Lecture Notes in Computer Science*, No. 993, pp. 25–39

Chegury Viana, F.A.; Iamin Kotinda, G.; Alves Rade, D. & Steffen, V. (2006). Can Ants Mechanical Engineering Systems?, *Proceeding of IEEE Congress on Evolutionary computation*, Vancouver, Canada

Chen, L.; Chen, J.; Qin, L. & Fan, J. (2004). A method for solving optimization problem in continuous space using improved ant colony algorithm, *Data Mining and Knowledge Management: Chinese Academy of Sciences*, Vol. 3327, pp. 61-70

Colorni, A.; Dorigo, M. & Maniezzo, V. (1992). Distributed optimization by ant colonies, *Proceedings in the First European Conference on Artificial Life, Elsevier Science Publisher*, pp. 134 -142

Deneubourg, J.L. & Goss, S. (1989). Collective patterns and decision-making. *Ethology, Ecology & Evolution*, No. 1, pp. 295-311

Deneubourg, J.L.; Pasteels, J.M. & Verhaeghe, J.C. (1983). Probabilistic behaviour in ants: a strategy of errors?, *Journal of Theoretical Biology*, No. 105, pp. 259-271

Dorigo, M. (1992). *Optimization, learning and natural algorithms*. PhD thesis, Dipartimento di Elettronica, Politecnico di Milano, Italia

Dorigo, M.; Bonabeau, E. & Theraulaz, G. (2000). Ant algorithms and stigmergy, *Proceeding in Future Generation Computer Systems*, Vol. 16, No. 8, pp. 851-871

Dorigo, M.; Di Caro, G. & Gambardella, L.M. (1999). Ant algorithms for discrete optimization, *Artificial Life*, Vol. 5, No. 2, pp. 137-172

Dorigo, M.; Maniezzo, V. & Colorni, A. (1996). Ant System: Optimization by a Colony of Cooperating Agents, *IEEE Transactions On Systems, Man., and Cybernetics – part B*, Vol. 26, No. 1, pp. 29-41

Gagné, C.; Gravel, M. & Price, W.L. (2002). Scheduling continuous casting of aluminium using a multiple objective ant colony optimization metaheuristic, *European journal of operational Research*, Vol. 143, pp. 218-229

Gagné, C.; Gravel, M. & Price, W.L. (2004). Optimisation multiobjectfs à l'aide d'un algorithme de colonie de fourmis, *INFOR*, Vol. 42, No.1, (February 2004), pp. 23–42

Goss, S.; Beckers, R.; Deneubourg, J.L.; Aron, S. & Pasteels, J.M. (1990). How trail laying and trail following can solve foraging problems for ant colonies, in: *Behavioural Mechanisms of Food Selection*, (Ed. R.N. Hughes), NATO ASI Series, G 20, Springer-Verlag, Berlin, pp. 661-678

Hamed, O.A.; Al-sofi, M.A.K.; Imam, M.; Mustafa, G.M.; Bamardouf, K. & Al-Washmi, H. (2001). Simulation of multistage flash desalination process, *Desalination*, Vol. 134, No. 1-3, pp. 195-203

Liang, Y.C. & Smith, A.E. (2004). An Ant Colony Optimization Algorithm for the Redundancy Allocation Problem (RAP), *IEEE transaction on reliability*, Vol. 53, No. 3, pp. 417-423

Ling, C.; Jie, S. & Ling, Q. (2002). A method for solving optimization problem in continuous space by using ant colony algorithm, *Journal of Software*, Vol. 13, No. 12, pp. 2317-2323

Mathur, M.; Karale Sachin, B.; Prive, S.; Jayaraman, V.K. & Kulkarni, B.D. (2000). Ant colony approach to continuous optimization, *Industrial and Engineering Chemistry Research*, Vol. 39, No. 10, pp. 3814–3822

Monmarché, N.; Venturini, G. & Slimane, M. (2000). On how Pachycondyla apicalis ants suggest a new search algorithm, *Future Generation Computer Systems*, Vol. 16, No. 8, pp. 937-946

Myers, R.H. & Montgomery, D.C. (2002). *Response surface methodology process and product optimization using designed experiments*, John Wiley & Sons, Inc., 2nd ed., New-York, USA

Nagesh Kumar, D. & Janga Raddy, M. (2006). Ant Colony Optimization for Multi-Purpose Reservoir Operation, *Water Resources Management*, Vol. 20, pp. 879–898

Pourtakdoust, S.H. & Nobahari, H. (2004). An extension of ant colony system to continuous optimization problems. *Proceedings in Ant Colony Optimization and Swarm Intelligence (ANTS) 4th International Workshop*, pp. 294–301

Roux, W.J.; Stander, N. & Haftka, R.T. (1998). Response surface approximations for structural optimization. *International Journal for Numerical Methods in Engineering*, Vol. 42, pp. 517–534

Socha, K. (2004). ACO for continuous and mixed-variable optimization. *Proceeding in Ant Colony Optimization and Swarm Intelligence (ANTS) 4th International Workshop*, in: *Lecture Notes in Computer Science*, Vol. 3172, pp. 25-36

Socha, K. & Dorigo, M. (2006). Ant colony optimization for continuous domains, *European journal of operational research*, Vol. 185, No. 3, pp. 1155-1173

Stander, N. (2001). The successive response surface method applied to sheet-metal forming, *Proceedings of the first MIT conference on computational fluid and solid mechanics*, pp. 481-485, June 12–15, 2001

Sun, H. & Lee, S. (2005). Response surface approach to aerodynamic optimization design of helicopter rotor blade, *International journal for numerical methods in engineering*, pp. 125–142

Zhang, T.; Rahman, S. & Choi, K.K. (2002). A hybrid method using response surface and pattern search for derivative-free design optimization, in: *Mang HA, Rammerstorfer FG, Eberhardsteiner J.*, (Ed.), *Proceedings of the fifth world congress on computational mechanics*, Vienna, Austria, July 7–12, 2002

Zhao, J.H.; Zhaoheng, L. & Dao, T.M. (2007). Reliability optimization using multiobjective ant colony system approaches, *Reliability Engineering and System Safety*, Vol. 92, No. 1, pp. 109-120

2

Particle Reduction at Metal Deposition Process in Wafer Fabrication

Faieza Abdul Aziz, Izham Hazizi Ahmad,
Norzima Zulkifli and Rosnah Mohd. Yusuff
Universiti Putra Malaysia
Malaysia

1. Introduction

Metal Deposition or metallization process is one of the processes in fabricating a wafer. A wafer is a thin slice of semiconductor material, such as a silicon crystal, used in the fabrication of integrated circuits and other micro-devices. Due to the nature on the process, it creates lot of particles, which would impact the next process if it were not removed. Particle deposition on the wafer surface can cause the circuit to malfunction; leading to a loss of yield. Cleaning process needs to be done after metal deposition process in order to remove the particles

Metal deposition, which has been constructed by several metal layers, allows the flow of current between interconnections. Each metal layers consist of three types of metal films such as Ion Metal Plasma Titanium (IMP Ti), Titanium Nitride (TiN) and Aluminum. The metal deposition started after the wafer has completed the "Tungsten Chemical Mechanical Polishing" (CMP) process.

Metal layers are deposited on the wafer to form conductive pathways. The most common metals include aluminium, nickel, chromium, gold, germanium, copper, silver, titanium, tungsten, platinum and tantalum. Selected metal alloy also may be used. The metal layer is shown in Figure 1 and the interconnection between metal layers is shown in Figure 2.

The deposited metal(s) offers special functionality to the substrate. Typically, the metal aqueous solution is employed for the wet metal deposition process due to the consideration of its low cost and operation safety.

Metallization is often accomplished with a vacuum deposition technique. The most common deposition processes include filament evaporation, electron- beam evaporation, flash evaporation, induction evaporation and sputtering. There are also two types of wet metal deposition processes – electrolytic and electro-less plating.

Sputtering and evaporation are well established as the two most important methods for the deposition of thin films. Although the earliest experiments with both of these deposition techniques can be traced to the same decade of the nineteenth century (Grove, 1852; Faraday, 1857), up until the late 1960s evaporation was clearly the preferred film-deposition technique, owing to its higher deposition rates and general applicability to all types of materials. Subsequently, the popularity of sputter deposition grew rapidly because of the need to fabricate thin films with good uniformity and good adhesion to the substrate surface (demand driven by the microelectronics industry) as well as the introduction of radio-frequency (RF) and magnetron sputtering variants.

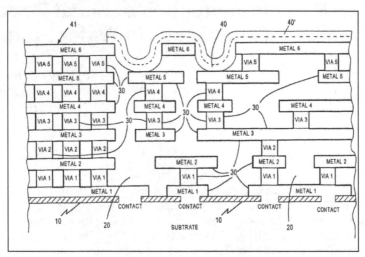

Fig. 1. The Metal layers

Fig. 2. The interconnection between metal layers

In this chapter, a thorough investigation was carried out to improve shut down event problem at Metal Deposition process during wafer fabrication. Particle contamination on wafer surface can cause the circuit to malfunction and leading to machine shut down. Data of shutdown event versus sputter target life showed that the rate of machine shutdown increased by the increment of sputter target life. The sputter target life was further investigated to determine the appropriate sputter target life to be used in order to avoid particles generation during metal deposition process.

2. Particles

Particles can be defined as "suspension of solid or liquid mass in air". Particles can originate from a variety of sources and possess a range of morphological, chemical, physical and thermodynamic properties. The particles could be combustion generated, photo-chemically produced, salt particles from sea spray or even soil-like particles from re-suspended dust. Particles may be liquid; solid or could even be a solid core surrounded by liquid.

Particles are represented by a broad class of chemically and physically diverse substances. Particles can be described by size, formation mechanism, origin, chemical composition, atmospheric behavior and method of measurement. The concentration of particles in the air varies across space and time, and is related to the source of the particles and the transformations that occur in the atmosphere. Some of the more generalized characterization of particles is:

i.   Primary and secondary particles: A primary particle is a particle introduced into the air in solid or liquid form, while a secondary particle is formed in the air by gas-to-particle conversion of oxidation products of emitted precursors.

ii. Particle characterization as per size: Particle can be classified into discrete size categories spanning several orders of magnitude, with inhalable particles falling into the following general size fractions- PM_{10} (equal to and less than 10 micrometre (μm) in aerodynamic diameter), $PM_{2.5-10}$ (greater than 2.5 μm but equal to or less than 10 μm), $PM_{2.5}$ (2.5 μm or less), and ultra fine (less than 0.1 μm).

iii. Particle characterization depending on requirements of study: Some of the particle components/ parameters of interest to health, ecological, or radiative effects; for source apportionment studies; or for air quality modeling evaluation studies are particle number, particle surface area, particle size distribution, particle mass, particle refractory index (real and imaginary), particle density and particle size change with density, ionic composition (sulphate, nitrate, ammonium), chemical composition, proportion of organic and elemental carbon, presence of transition metals crustal elements and bioaerosols

2.1 Particle contamination

Particle contamination can be defined as the act or process of contaminating by particulates. Particle contamination is problematic for many industries. They can appear unexpectedly mixed in solids, liquids and gases. Particles can be from many sources i.e.- metals, biological (skin, hair etc), polymers, building dusts etc. They all have different characteristics and properties such as shape, size and chemistry, which assist in identification. Scanning Electron Microscope (SEM) and Energy-dispersive X-ray spectroscopy (EDX) coupled with optical microscopy provides a powerful machine for unambiguously identifying such particles. The technique is frequently coupled with Fourier transform infrared spectroscopy (FTIR) when identifying the source of organic contamination (Stephen, 2010).

2.1.1 Particle contamination in semiconductor

Deposition of aerosol particles on semiconductor wafers is a serious problem in the manufacturing of integrated circuits. Particle deposition on the wafer surface can cause the circuit to malfunction, leading to a loss of yield. With the circuit feature approaching 1 μm in size of one-megabit memory chips, particle control is becoming increasingly more important (Benjamin et al., 1987). Particle contamination during vacuum processing also has a significant impact in Very Large Scale Integration (VLSI) process yield (Martin, 1989) and has motivated most manufacturers to adopt particle control methods base on sampling inspection.

According to Bates (2000), semiconductor memory chips are very sensitive to the particles because the circuitry is so small. In a typical clean room manufacturing environment, particles are deposited on the wafer surface by sedimentation, diffusion, and/ or electrostatic attraction. Sedimentation usually occurs for large particles, particularly those larger than 1 μm in diameter, whereas diffusion occurs for small particles below 0.1 μm in diameter.

In the intermediate size range, both sedimentation and diffusion may occur and must be considered. When particles are electrically charged, enhanced deposition can take place. The rate of particle deposition on a wafer surface depends on both the size of the particle and their electrical charge. In addition, the deposition rate is also influenced by the airflow around the wafer, which in turn are affected by the size of the wafer, the airflow velocity, and the orientation of the wafer, with respect to the airflow. Although the mechanisms of particle deposition on semiconductor wafers are reasonably well understood and approximate calculations have been made (Cooper, 1986; Hamberg, 1985), no detailed quantitative calculation has been presented.

2.1.2 Particle contamination in wafer processing

As the chip density increases and semiconductor devices shrink, the quality of fabrication becomes more crucial. The composition, structure, and stability of deposited films must be carefully controlled and the reduction of particulate contamination in particular becomes increasingly crucial as device sizes shrink and densities increase. As the devices grow smaller, they become more sensitive to particulate contamination, and a contaminant particle size that was once considered acceptable may now be a fatal defect. Voids, dislocations, short circuits, or open circuits may be caused by the presence of particles during deposition or etching of thin films. Yield and performance reliability of microelectronic devices may be affected by the mentioned defects (Alfred, 2001).

Often, the process gases will react and deposit material on other surfaces in the reactor besides the substrate. The walls of the processing chambers may be coated with various materials deposited during processing, and mechanical and thermal stresses may cause these materials to flake and become dislodged, generating contaminated particles. In processing steps that use plasma, many ions, electrons, radicals, and other chemical "fragments" are generated. These may combine to form particles that eventually deposit on the substrate or on the walls of the reactor (Alfred, 2001). Particulate contamination also may be introduced by other sources, such as during wafer transfer operations and backstream contamination from the pumping system used to evacuate the processing chamber.

In plasma processing, contaminated particles typically become trapped in the chamber, between plasma sheath adjacent to the wafer and plasma glow region. These particles pose a significant risk of contamination, particularly at the end of plasma processing, when the power that sustains the plasma is switched off. In many plasma-processing apparatuses, a focus ring is disposed above and at the circumference of the wafer to enhance uniformity of processing by controlling the flow of active plasma species to the wafer, such as during a plasma etch process. The focus ring, and the associated wafer clamping mechanism, tends to inhibit removal of the trapped particles by gas. Thus, there is a need to provide a reliable and inexpensive process to remove such particles from the wafer-processing chamber (Alfred, 2001).

Similarly, in chemical vapor deposition and etching, material tends to deposit on various parts of the apparatus, such as the susceptor, the showerhead, and the walls of the reactor, as the by-products of the process condenses and accumulates. Mechanical stresses may cause the deposited material to flake and become dislodged. These mechanical stresses are often caused by wafer transfer operations, but may also be caused by abrupt pressure changes induced by switching gas flow on and off and by turbulence in gas flow. Thus, process by-products at the end of the processing stage must be flushed from the chamber to prevent them from condensing and accumulating inside the chamber.

Typically, the flow of the processing gas is shut off at the end of a processing stage, whereupon the pressure in the chamber rapidly falls to zero as the vacuum pump continues to run. Idle purge may be used; in which purge gas is introduced into the chamber at intervals while no processing is taking place. Nonetheless, pressure spikes occur with the cycling of gas flow, causing disruption of particles, which may then contaminate the wafer surface. This limits the particle reduction benefits from the idle purge. A large portion of device defects is caused by particles disrupted by pressure change during wafer loading and moisture on the pre-processed wafer surface (Alfred, 2001).

Three types of particle contamination can be defined, which are under the deposited film as shown in Figure 3, in the deposited surface as shown in Figure 4 and deposited Film as shown in Figure 5. Particle under the deposited film will cause the surface of the wafer to become dirty. The particle may come from the previous process. Particle in the deposited surface will cause gas phase nucleation, leaks into the system, contamination in gas source/flow lines and sputter off walls. The particle may come from the gas phase nucleation, system leak or contaminated gas line. Particle on the deposited film will cause film build-up on the chamber walls. The source may come form the process chamber or from the wafer handling.

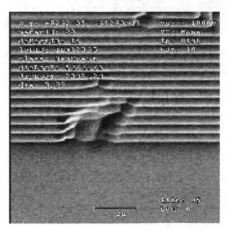

Fig. 3. Particle under the deposited Film

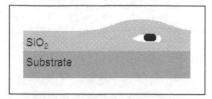

Fig. 4. Particle in the deposited surface

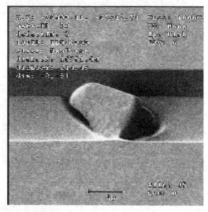

Fig. 5. Particle on the deposited Film

Example of TiN particle transformation is shown in Figure 6. From this figure, the particle was dropped on the wafer's surface. The source of the particle may come either form previous process or from current process. After the deposition process done, the particle will be covered underneath the metal layer, which cause the damage of the interconnection.

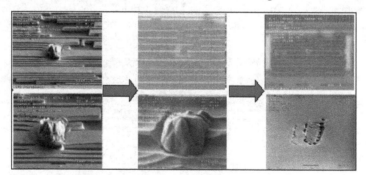

Fig. 6. The transformation of TiN particle

The particles entrained in the load lock air volume by turbulence during pumping are either carried in through the handling of the wafer, generated within the camber from causes such as wear or residual from previous pumping and venting cycles. Particles are removed as they are drawn out during pumping or as they are carried out of the surface of the wafers. Additional particles may bind to the walls of the chamber or machining to tightly that they are agitated free by subsequent pump/ vent cycles (Peter Bordon, 1990).

An equilibrium background level is reached because the number of particles carried out by pumping and deposition on the wafer surface is proportional to the number of particles entrained into the gas volume. For example, if the number of particles entrained doubles, twice as many particles land on the wafer and twice as many flow out the pump line (Bordon, 1990). The effectiveness of these mechanisms has long been recognized. For example, it is a common practice to pump/ vent clean process chambers in high- current ion implanters and other process machines or to run getter wafers after a chamber has been contaminated.

Low levels of particulate contamination can be obtained in process gas systems by using careful system design, high-quality compatible materials, minimum dead legs and leak rates, careful start-up and operating procedures, etc. Low particle levels can also be obtained in gas cylinders through careful selection of cylinder materials, surface treatment and preparation, and through close attention to gas fill system design and operation (Hart, et al., 1994).

Particle levels in flowing gas systems may be steady or (as in machine vent lines) cyclic over time. In machine feed lines, the gas is usually well mixed and particles are uniformly distributed. However, particle levels in gas cylinders can vary by orders of magnitude over time due to such effects as liquid boiling, gravitational settling, and diffusion to internal surfaces. Such effects may also produce non-uniform particle distributions, including stratification, in gas cylinders (Hart, et al., 1995). Levels of suspended particles in filled cylinders can be measured with a high-pressure Optical Particle Counters (OPC). Data obtained directly from cylinders show that careful attention to quality can result in low cylinder particle concentrations.

Cylinder and bulk gases are frequently reduced in pressure with an automatic regulator before entering the flowing distribution system. Automatic regulators may produce increased particle levels (through regulator shedding, impurity nucleation, and condensational droplet formation) that are sometimes followed by system corrosion (Chowdhury, 1997) or suspended nonvolatile residue formation. Gases are therefore filtered after pressure reduction and before entering the distribution system. Ceramic, metal, or polymer membrane filters are selected for compatibility with the process gas. Such filters can produce a low particle level as well as a low degree of variability in contamination over time.

CNC data for particles as small as $0.003\mu m$ in O2 and H2 can also be obtained using an inert gas CNC with a special sample dilution device developed by Air Products (McDermott, 1997). These data showed that membrane filters can be used to produce high-cleanliness gases to $0.003\mu m$ in large-volume gas systems. Well-designed distribution systems should contribute a minimum of additional particulate contamination to the flowing gas.

As the particle may impact the wafer quality, which result in wafer scrap, corrective and preventive action must be made immediately to stop the particle contamination from becoming catastrophic. Thus, a systematic problem solving method is needed to solve the issue.

2.2 Particle failure

In this step, the types of particle failure were studied. Data from Daily Particle Qualification process was analyzed. TiN particle have highest standard deviation ~4.0 compared to Aluminium (Al) and Ion Metal Plasma Titanium (IMP Ti) as shown in Figure 7. This showed that TiN particle performed the most inconsistent compared to Al and Imp Ti particle in the Metal Deposition process. Data for each films particle qualification was obtained and then Pareto Analysis was made. From the Pareto Chart, TiN defect has the highest failure rate compare to other films as shown in Figure 8.

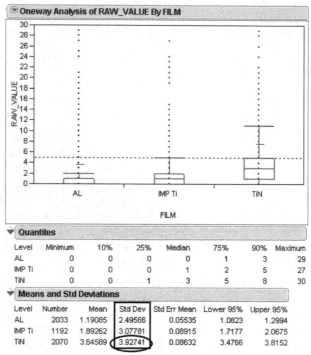

Fig. 7. Comparison between Al, IMP Ti and TiN

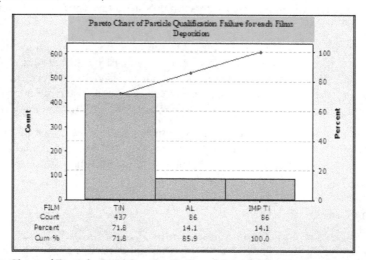

Fig. 8. Pareto Chart of Particle Qualification Failure for each Films Deposition

2.3 Chamber configuration and wafer processing sequence

Chamber configuration for metal deposition machine is shown in Figure 9. From below chamber configuration, the wafers which inside the cassette are placed at cassette Load lock, which consist of Load lock A (LLA) and Load lock B (LLB). Wafer will pass through form Buffer camber to transfer chamber in Chamber A. Then, wafer will be cooled down in Chamber B. At the end of the process, wafers will be vented to Atmosphere condition in Load lock A or Load lock B, depending to which Load lock the wafers origin.

Wafers in a production pod will be pumped down to vacuum condition from atmospheric pressure in load lock A or load lock B, depending on where the lot is placed. Degas and notch alignment occurred in Chamber E and F. The deposition process begins with IMP Ti deposition in Chamber C. The wafers will move into the transfer chamber in Chamber A. Metal deposition will occur in Chambers 1, 2, 3 and 4. Depending on the application, multiple metal films can be stacked without breaking the vacuum. After the deposition process, wafer will be cooled down in Chamber B and vented back into the atmosphere in the production pod at Load lock A or B.

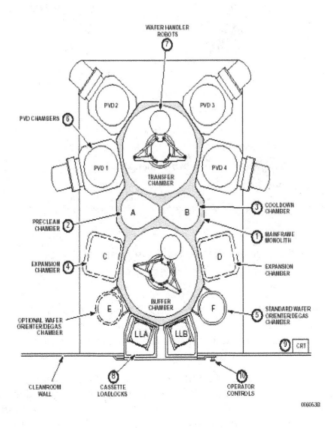

Fig. 9. Chamber Configuration of the Metal Deposition Machine

3. Cause and effect analysis diagram

A case study has been conducted in one wafer fabrication company. Downtime reduction at Metal Deposition process is being focused in this work. In average, three cases of Metal Deposition machines have been shutdown every week. Shutdown criteria is based on more than "10-area count per wafer" or well known as "adders". The machine will be shut down if the post scan result shows particle increase more than 10 adders.

Brainstorming session has been done with the team members and root causes have been identified and classified under six main sectors, which are machines, material, methods, measurements, environment and personnel. Figure 10 is the fish bone diagram for cause and effect analysis.

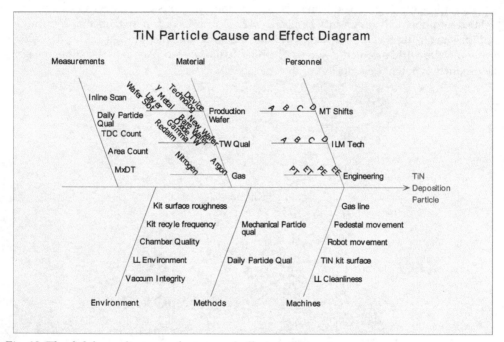

Fig. 10. The fish bone diagram of cause and effect analysis

3.1 Measurement systems

Particle measurement is performed by using the SP1 machine as shown in Figure 11. SP1 is the machine, which is measuring the particle by scanning and counting the existing particle on the wafer. Besides that, SP1 also is capable to show the wafer map, which can tell either the particles are clustered on the wafer, saturated, mild signature and others.

Percentage of Gage Repeatability and Reproducibility (GRnR) was done on SP1 in the production floor, which is used for particle scanning purpose. Gage RnR study is conducted to determine the measurement system variability in term of Repeatability and Reproducibility. TiN particle count is our KPOV that causing machine shutdown by ILM due to metal deposition particle existence in the machine.

Based on the GRnR study, the result proves that the SP1 is capable to measure the TiN Dep particle counts cine the total GRnR is less than 30%. The result of GRnR study is shown in Figure 12.

Fig. 11. The SP1 machine which measure the particle

```
Gage R&R

                                %Contribution
Source                 VarComp   (of VarComp)
Total Gage R&R         0.67500         6.88
  Repeatability        0.38333         3.91
  Reproducibility      0.29167         2.97
    Day                0.14722         1.50
    Day*Slot           0.14444         1.47
Part-To-Part           9.13519        93.12
Total Variation        9.81019       100.00


                                  Study Var  %Study Var
Source               StdDev (SD)   (6 * SD)    (%SV)
Total Gage R&R          0.82158      4.9295     26.23
  Repeatability         0.61914      3.7148     19.77
  Reproducibility       0.54006      3.2404     17.24
    Day                 0.38370      2.3022     12.25
    Day*Slot            0.38006      2.2804     12.13
Part-To-Part            3.02245     18.1347     96.50
Total Variation         3.13212     18.7927    100.00


Number of Distinct Categories = 5
```

Fig. 12. The result of GRnR study for SP1

3.2 In-line monitoring and systematic machine excursion monitoring

There are three methods of inspecting and measuring the particle, which are using production wafers through Systematic Machine Excursion Monitoring (STEM), production wafer that went to In-line monitoring process (ILM) flow and test wafers which is being used during machine qualification process.

In- line monitoring (ILM) is a process to detect any defect in real time. It is done in many ways, such as in line inspection, upon request from user, from production lots which go to ILM flow and also Systematic Machine Excursion Monitoring (STEM) lots. Machine-related defect excursions are controlled by systematically checking process machines. Production wafers are being used for STEM purpose. Each of Metal Deposition machine need to do STEM activity once every two days. STEM is a process where the lot which is already completely processed from one machine, will be held for ILM scan. The scan is done to check for any defects that may be caused by the processing machine at previous process. STEM will provide faster detection and containment of the defect excursion. All major process machines are monitored in a systematic manner.

For STEM activity, Manufacturing Technician will hold the lot for ILM Technician after run through the metal deposition machine. ILM Technician will scan four wafers/ machine using Complus or AIT machine. If the scan result shows particle signature and above the control limit (more than 10 counts), they will shutdown the whole machine and the machine owner need to verify the shutdown prior to release the machine back to production.

For production wafers, there will be about 30% of the WIP will go to ILM inspection step. This is the random sampling in line scanning that has been designed to detect any defects along the process of fabricating the wafers from first process until end on the process. It has been designed in the process flow, where lots that are needed for this sampling will have ILM inspection flow compare to the other 70% of the lots that do not have ILM flow. Lots that have ILM flow will arrive at ILM inspection step after completing metal deposition process. ILM technician will scan the lot and if found particle and above the control limit (more than 10 counts), machine will be shutdown and same verification need to be done prior to release the machine back to manufacturing.

For qualification process, bare wafers or known as test wafers is used to check the machine's condition and performance. Qualification process is done based on schedule. Basically, every metal deposition machines need to perform qualification process once everyday. This is to ensure the machine is fit to run and not causing any defect later. Qualification process is carried out by manufacturing technician using SP1.

The qualification process is started with the pre particle measurement. Qualification wafer (bare wafer) will be selected and pre particle measurement is done using SP1. After pre-measurement is completed, the wafer will go inside the machine and process chamber for machine and chamber qualification purpose. After the process is completed, the wafer is again brought to SP1 for post particle measurement. The differences between pre particle value and post particle value will determine the machine and chamber's condition. For qualification process, the control limit is tightened to five count only. If particle is found more than five count, chamber will be shut down and pending verification from machine owner is needed prior to release the machine back to production.

Example of the pre particle and post particle measurement is shown in Figure 13. In Figure 13, two wafer maps were shown, which are pre particle wafer map and post particle wafer map. In Pre Particle wafer map, two particle counts were detected as circled. In post particle wafer map, four particle counts were detected. Two count were the existing particles and the other two were new particles, which were detected during post particle measurement. From Figure 13, the adders were two counts (post particle value- pre particle value).

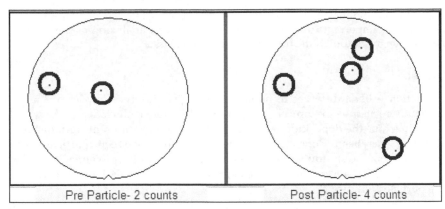

| Pre Particle- 2 counts | Post Particle- 4 counts |

Fig. 13. Pre Particle and Post Particle Wafer Map

3.3 Possible root cause

From Ishikawa Diagram, possible root causes will be screened out to get the actual causes. From actual causes, potential corrective and preventive actions will be determined and implemented.

3.4 Personnel

There are four shifts in the studied company, which are shift A, B, C and D. Manufacturing technician (MT) for every shift is responsible to perform the daily qualification job daily. Since there are four shifts running in the production, the level of experiences between shifts to shift differs. The level of experiences of the MT is very important since they need to perform the qualification process. Experienced MT will know and easily catch the particle issue inside the chamber by looking at the qualification result, but less experienced MT may take some times. Study has been done to check the Manufacturing Technician's efficiency. Their year of services and also certification were referred. The data were obtained from Human Resources Certification Record. Level of certification is from one to three. Level one is the minimum certification level, while level three is the maximum level of certification.

Study has been done to check the ILM Technician's efficiency. Their year of services and also certification were referred. The data were obtained from Human Resources Certification Record.

Process technician, process engineer, equipment technician and equipment engineer also play their roles during machine shutdown. When machine detected unwanted particle and need to be shutdown, process technician normally will follow the Out of Control Action Plan (OCAP) in order to release the machine back to manufacturing group as soon as possible. Process engineer also will take a look on the issue and do analysis and then come out with the release plan. Equipment technicians and engineers need to ensure proper maintenances job been carried out as per checklist. This is to ensure the cleanliness of the machine after Preventive Maintenance (PM) was done. Study has been done to check the Equipment and Process Technician's efficiency. Their year of services and also certification were referred. The data were also obtained from Human Resources Certification Record.

Beside MT, In-Line monitoring (ILM) technician also play big responsibility to determine the particle rate. It is important to have a proper scanning and analyzing of the STEM lot, so that the decision to shutdown the machine is base on real issue.

3.5 Material

For production wafers, different technologies will give different impact of the particle. This is mainly related to the process recipes, which different devices will have different process recipe, thus the deposition rate and thickness will be different from one device to others. Study has been done to see the relationship between particle issue and technologies. From the shutdown event, list of lots that have been scanned was obtained. From the list, product technologies were segregated and the relationship between them with the particle is studied.

Beside the technology, the metal layers also have impact to the particle issue since more metal layers means more times the lot will go to metal deposition process and the chances for expose to particle issue is more. Example for lot with four layer metal will go four times metal deposition process compare lot with five metal layers, will go five times metal deposition process. Study has been done to see the relationship between particle issue and the metal layers. From the same list from shutdown event, metal layers were obtained to see if there is any relationship between metal layers and particle.

Test wafer also have some impact to the particle issue. For new test wafer, the performance is better compared to wafers that sent to rework and reused. This is because the rework wafer normally will have remaining particle, which can not be removed due to saturated at the surface of the wafer and needs stronger cleaning recipe to remove them. Brand new wafers normally will have a lot less particle. In this study, the incoming particle for 50 lots of new test wafers was measured using SP1 to get the potential incoming particle. From here, any existing particle from test wafers itself that may contaminate the process chamber later during qualification process can be seen.

3.6 Method and measurement

Correct methods, which are used during both particle and mechanical qualification, were studied and observed. Judgment was made base on observation across all four shifts on the procedures during the qualification process.

The particle measurement is done based on In-line scan and during particle qualification process. For inline particle scanning, it is done after the lot has completed the metal deposition process. The job of in-line scanning is known as Systematic Machine Excursion Monitoring (STEM), which been done once in two days. Lot will be on hold for in line monitoring (ILM) scan.

Four wafers will be scanned for each machine to check for particle performance. The wafers will be scanned using Scanning Electron Microscope (SEM) machine. If particle signature exists, ILM personnel will notify the Metal Deposition machine owner to check for the machine's health. If the particle level exceeds the limit, which is more than 10 particle counts, the machine will be shut down and need to follow procedures in order to bring the machine back up to the production.

For production lots that go to ILM flow, the lot will be scanned for scratches and particle. If scratches or particle are found to exceed the limit, which is more than 10 particle counts, machine will be shutdown and need to follow the procedure as well. Wafers will be scanned using Complus or AIT machine also.

Qualification process is a process to check for the machine's performance, so that it always performs same as the baseline. One of the important factors in qualification is the particle performance. Particle value is measured based on the different value between pre particle measurement and post particle measurement. Differences of both values will determine the particle existence inside the machine. If particle count is more than five area count/ wafers, machine will be shutdown and need to be followed up by machine's owner before release back to production.

3.7 Machine

Machine is the main focus of the particle issue. This is due to the mechanical movement such as pedestal and robot movement inside the machine that can generate particle. Beside that, gas line also can create particle.

Load lock cleanliness is also very important since this is the place where the lot is transferred into the machine from its base. Load lock is a chamber that is used to interface a wafer between air pressure and the vacuum process chamber. According to Borden (Borden, 1988), Wu (Wu, et al., 1989) and Chen (Chen, et al., 1989), in the absence of a water aerosol, the dominant source of wafer contamination is the agitation of particles during the pumping (venting) of the entry (exit) load lock.

In this study, 100 lots were selected to check the particle level in the cassette. Since wafers are inside the vacuum state inside the cassette, particle inside the cassette need to be measured. It was measured using mini- environment tester. The cassettes were opened in Wafer Start room and the particle was measured for all the 100 lots. The particle count that obtained from the testing is captured.

To study for particle during wafer handling and robot movement, mechanical qualification process was carried out. Before it was done, chamber was cleaned first to eliminate the potential source of particle coming from process chamber. One lot, which consists of five test wafers, was selected. Wafers inside the cassette were arranged in slot 5, 10, 15, 20 and 25. Pre particle measurements were obtained for all the five wafers. The lot then was vented inside the load lock and also into the deposition module. Without running the deposition process, the lot was moved out back into the load lock and cassette. Post particle measurement was done to check for the adders. This cycle is repeated for 10 times for the entire machine and data is captured and analyzed.

Particle in gas line also was focused in this study. Particle in gas line is measured by referring to the data that is obtained from the particle sensor. The particle sensor is mounted at the gas line as shown in Figure 14. This is to ensure any particle in the gas line can be detected and the amount of particle entered to the process chamber can be monitored and recorded.

Sputter target also been studied to check the correlation between sputter target life and also shutdown. Example of sputter target that been used inside metal deposition machine is shown on Figure 15. The event of shutdown and the usage of target life is captured and analyzed.

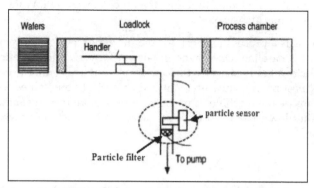

Fig. 14. Particle Sensor that Mounted at Gas Line

Fig. 15. Example of Sputter Target

3.8 Environment

The machine environment also has been studied to see any contribution to shutdown event. Load lock environment has been checked. Airborne particle measurement was conducted over metal deposition machines to collect the particle count. One hundred lots were prepared in this study. Wafers inside the cassette were arranged in slot 5, 10, 15, 20 and 25. Pre particle measurements were obtained for all the five wafers. The lot then was vented inside the load lock and left for five minutes. After five minutes, lot was moved out back from load lock move into cassette. Post particle measurement was done to check for the adders.

4. Results and discussion

From the fish bone diagram, all the possible causes have been screened out and verified to find the actual true causes. From actual true causes, corrective actions and preventive actions will be defined, identified and will be implemented to eliminate the particle issue.

4.1 Personnel

Verification have been made to people who directly working at metal deposition machine and related to the shutdown. Summary of the possible causes, which related to personnel, is shown in table 1.

	Causes	Verification & Validation Process	Result	True/ False
Personnel	Manufacturing Technician (MT) in shift	1. To check the level of experiences of the MT 2. To check the capability of MT to perform qualification job correctly	1. Base on the study, all shift have dedicated MT > 2 years of experience to handle Metal Deposition tool 2. All the Manufacturing Technicians also capable to perform qualification job corectly	FALSE
	In Line Monitoring (ILM) Technician in Shift	To check the capability of ILM Technician to handle to scanning tool and catch the particle	Base on the study, all shift ILM Technician also having > 2 years of experience and capable to handle the scanning tool	FALSE
	Engineering Personnel (Process/ Equipment)	1. To check the capability of shift Equipment Technician (ET) and Equipment Engineer (EE) to perform Preventive Maintenance (PM) job efficiently. 2. To check the capability of Process Technician (PT) and Process Engineer (PE) to follow up on the ILM Shutdown issue to avoid re-shutdown	1. All shift ET/ PT and Engineers (PE/ EE) are well trained and have experiences > 3 years in average. 2. Equipment & Process team have their own checklist to be followed and verified by Section Head.	FALSE

Table 1. Summary of verify possible root causes related to personnel

4.1.1 Manufacturing technician (MT)

A validation and verification have been done to check the level of experiences of the MT and also the capability of MT to perform qualification job correctly. From the study, all shifts have dedicated MT more than two years of experience to handle Metal Deposition machine due to the criticality of metal deposition process. All the Manufacturing Technicians are also capable to perform qualification job correctly based on the checklist. The dedicated MT is summarized in table 2.

Shift	Person in Charge	Date Join	Years of Experiences	Certification level
A	1. NEDUMARAN A/L MEGAWARNAM	2004	> 5 years	3
	2. LIYANA HANIM BINTI AKBAR	2006	> 3 years	3
	3. MUHAMAD TERMIZI BIN AHMAD TAJUDIN	2006	> 3 years	3
B	1. ROZI BIN MD HASSAN	2003	> 6 years	3
	2. NOOR JANNAH BINTI MAHADZIR	2007	> 2 years	3
	3. ANUAR BIN MAT ISA @ ABDUL AZIZ	2007	> 2 years	3
C	1. MOHD ASRIZAL BIN AHMAD	2003	> 6 years	3
	2. KASMINI BINTI TUKOL	2007	> 2 years	3
	3. MAIMUNAH BINTI HASHIM	2007	> 2 years	3
D	1. MOHD YUSRI BIN YUSOF	2007	> 2 years	3
	2. MOHD IZHAM BIN MOHD IZHAR	2004	> 5 years	3
	3. ZAIDA BINTI AHMAD	2007	> 2 years	3

Table 2. Manufacturing Technicians in-charged of Metal Deposition Machine

4.1.2 Inline monitoring (ILM) technician

Based on the study, all shift ILM Technicians are also having more than two years of experience and capable to handle the scanning machine and captured defect images. The shift ILM Technician is summarized in table 3. Conclusion can be made that all the MT who handle the medal deposition machines are capable to perform qualification process and mistake that can lead to particle generation is almost zero.

Shift	Person in Charge	Date Join	Years of Experiences	Certification level
A	REDZUAAN BIN ABDUL RAHIM	2002	>7 years	3
	BALAKRISNAN A/L A.MUNIANDI	2007	> 2 years	3
B	RUZAINI B. ADZHA	2006	> 3 years	3
	CHAREN A/L KHAN	2007	> 2 years	3
C	SUNTHARA MURTHI S/O RAMAN	2003	> 6 years	3
	ERUAN BIN ABU SEMAN	2007	> 2 years	3
D	VASANTHAN A/L VELOO	2007	> 2 years	3
	IBRAHIM BIN IDRIS	2005	> 4 years	3

Table 3. In-line monitoring (ILM) Shift Technicians

4.1.3 Engineering personnel (process/ equipment)

Verification and validation made to check the capability of shift Equipment Technician (ET) and Equipment Engineer (EE) to perform Preventive Maintenance (PM) job efficiently. Also validation made on the capability of Process Technician (PT) and Process Engineer (PE) to follow up on the ILM Shutdown issue to avoid re-shutdown due to incorrect qualification job done prior releasing machine during shutdown.

The summary of PT is shown in table 4 and summary of ET is shown in table 5. Based on the verification, all shift ET/ PT and Engineers (PE/ EE) are well trained and having experiences to perform their job efficiently. Equipment and Process team have their own checklist to be followed and verified by Section Head during performing PM activities and also releasing the machine from shutdown.

Shift	Process Technician	Date Join	Years of Experiences	Certification level
A	NOR AZELINA BINTI ISMAIL	2002	>7 years	3
	HARYANI BINTI ABDULLAH	2007	> 2 years	3
B	NORMALA BINTI NAPIAH	2006	> 3 years	3
	MOHD SYUKRI BIN CHE HASSAN	2007	> 2 years	3
C	KHAIRUL ANWAR BIN ABU BAKAR	2003	> 6 years	3
	CANITTHA A/P IEKIN	2007	> 2 years	3
D	PUTERI SURINAEDAYU BINTI MEGAT ISMAIL	2007	> 2 years	3
	NOR ADILA BINTI ABDUL RASHID	2005	> 4 years	3

Table 4. Shift Process Technicians for Thin Film Metal Module

Conclusion can be made that PT who work at metal deposition process is capable to perform machine recovery as per procedure during the event of ILM shutdown. Particle generation during recovery or re-occurrence of shut down due to wrong recovery is zero. Conclusion also can be made that all ET who working with metal deposition machines are capable to perform preventive maintenance jobs effectively and mistake that can lead to particle generation is almost zero.

Shift	Equipment Technician	Date Join	Years of Experiences	Certification level
A	MOHAMMAD RIDZAL BIN ABDULLAH	2002	>7 years	3
A	MOHD KHADAFI BIN MAHAMOD	2007	> 2 years	3
B	AYUB BIN AHMAD	2006	> 3 years	3
B	FRANCIS SELVAN A/L SINNAYAH	2007	> 2 years	3
C	MOHD ABDUL WAFI BIN AHMAD NADZIR	2003	> 6 years	3
C	MOHD AZHUZAIRI BIN ABDULL AZIZ	2007	> 2 years	3
D	KHAIRUL HYFNI BIN NOORDIN	2007	> 2 years	3
D	MOHD FAHMI BIN MOHD TAIB	2005	> 4 years	3

Table 5. Shift Equipment Technicians for Thin Film Metal Module

4.2 Material

Validation and verification were made in order to study the relationship between materials used in metal deposition process, with the shutdown rate related to particle, as shown in Table 6. The number of shutdown event (weekly) versus output (wafer move out from equipments) is shown in Figure 16. In average, three machines will be shutdown for every 27,900 wafer output from metal deposition process.

4.2.1 Relationship between shutdown event and output (weekly)

By using Minitab, correlation test between shutdown event and output has been conducted. The Pearson correlation between shutdown and output is 0.005, which means no correlations between both variables.

Regression analysis between shutdown event versus output was made using Minitab. The result of R square (R^2) is zero, which means no relation between shutdown event and output.

	Causes	Verification & Validation Process	Result	True/ False
Material	Production Wafers	To check the relationship between shutdown and moves	1. Correlations test between shutdown and moves had been done. The Pearson correlation between shutdown and move is 0.005, which means no correlations between both variables. 2. Regression analysis between shutdown versus Moves was made using Minitab as shown in Figure 4.3. From the result, the R square (R2) is zero, meaning no relation between shutdown and moves.	FALSE
		To check the relationship between shutdown and moves by technologies	1. Pearson Correlation test was done between shutdown and each of the moves by technologies as shown in Figure 4.6. 2. From the Pearson Correlation test, conclusion can be made that no correlation is exist between shutdown and technologies. 3. Base on the regression analysis, no significant relationship between the technologies and the shutdown event. The R2 also showed 0%, which meaning no relationship between the shutdown and technologies.	FALSE
		Relationship Between Shutdown and Technologies (STEM Failed)	1. From the Pearson Correlation shown in Figure 4.8, conclusion can be made that no strong correlation between shutdown and technologies. 2. Regression analysis was made between the shutdown and technologies as shown in Figure 4.9 and no strong evidence to conclude that the shutdown is influenced by the technology.	FALSE
		Relationship Between Shutdown and Metal Layers (LM)	1. From Figure 4.11, no strong evidence to say that the shutdown is influenced by the four layer metal lots. 2. From Regression analysis in Figure 4.12, also indicates no strong correlation between shutdown and layer metal.	FALSE
	Test Wafers	Relationship Between Incoming Particle from New Test Wafer and Shutdown	All the 50 lots of new test wafer were passed the incoming particle screening. From this study, conclusion can be made that the particle generation is not coming from the new test wafers that were used during qualification process.	FALSE

Table 6. Summary of possible root causes related to materials used in metal deposition

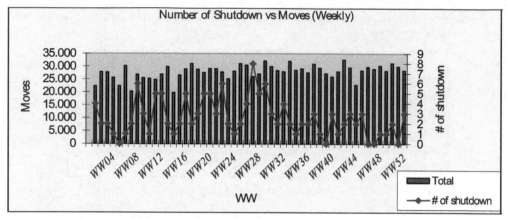

Fig. 16. Number of Shutdown Event versus Machines Output (Weekly)

4.2.2 Relationship between Incoming particle (from new test wafer) and shutdown event

Incoming particle screening was performed for 50 lots of new test wafers. Histogram was generated as shown in Figure 17. From the results, the mean is 1.18 with standard deviation of 1.24. Out of 50 lots of new test wafer that have been measured, 23 lots resulted in zero count of incoming particle, 6 lots showed one count of incoming particle, 10 lots showed two count and 11 lots showed three counts of incoming particle. Since the specification for the incoming particle is five count, all the 50 lots of new test wafer passed the incoming particle screening. From this study, conclusion can be made that the particle generation is not caused by the new test wafers that were used during qualification process

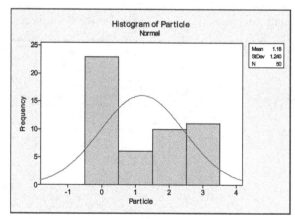

Fig. 17. Histogram for Incoming Particle in New Test Wafer

4.3 Method and measurement

Method of executing the qualification process were observed and summarized. Since the entire MT whose handle metal deposition machines were at level three, therefore they are considered as competent to perform the task efficiently. This is proved by the qualification data that is available in the spreadsheet and also in the CIM system. By looking at the shutdown trend, conclusion can be made that STEM is effective to detect particles that generated at Metal Deposition process.

4.4 Machines

The relationship between shutdown event and machine was studied and the summary of the result is shown in table 7.

Causes		Verification & Validation Process	Result	True/ False
Machine	Vacuum Cassette	To check the particles that may enter a cassette from the cleanroom	Particle and mechanical qualification was done to check the particle existance. Result was clean and no particle was found during wafer loading from production pod to the loadlock	FALSE
	Wafer Handler/ Robot movement	To check the particle that may be created during wafer transfer from vacuum cassette to deposition module.	Particle and mechanical qualification was done to check the particle existance. Found that particle was created during wafer loading from vacuum cassette to deposition module	TRUE
	Gas Line	To check the particle in the gas line	Base on the particle data which is obtained from the particle sensor, no partilce can escape through the particle filter.	FALSE
	Sputter Target	To check the relationship between Sputter Target Life with ILM Shutdown	Base on the correlation analysis, there is relation between Sputter Target Life and ILM Shutdown	TRUE

Table 7. Summary of verify possible root causes related to Metal Deposition Machine

4.4.1 Relationship between shutdown event and particle in vacuum cassette

Results of particle existence in vacuum cassette are shown in Figure 18. From the bar chart, 82 lots detected zero particle count, 13 lots showed particle with one count and five lots showed two counts of particle. Conclusion can be made that particle almost not exist and can be considered as negligible in vacuum cassette, as the production is running under clean room environment of class one category.

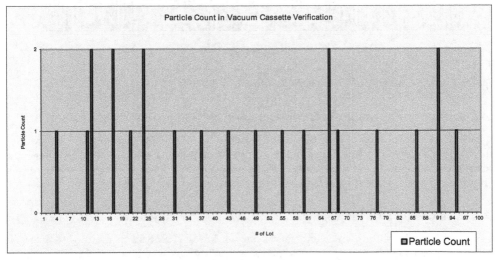

Fig. 18. Bar Chart for Particle Existence in Vacuum Cassette

4.4.2 Relationship between shutdown event and particle generation during mechanical movement

Results for mechanical qualification process are tabulated in table 8. From the data, bar chart was generated as Figure 19. From the bar chart, each of the machines showed particle generation inside the load lock. From the data, conclusion can be made that particles can be generated during wafer handling and mechanical movement.

Tool	Particle Count (adders) for mechanical Qualification Process for all Metal Depositon Tool									
	1st	2nd	3rd	4th	5th	6th	7th	8th	9th	10th
Metal Dep 01	8	1	5	11	2	1	5	5	9	4
Metal Dep 02	12	3	8	2	2	4	12	5	9	8
Metal Dep 03	3	7	7	14	9	1	7	11	1	7
Metal Dep 04	15	1	7	2	10	2	5	8	2	9
Metal Dep 05	6	7	14	7	5	4	12	6	12	8
Metal Dep 06	3	2	1	3	5	13	9	3	7	11
Metal Dep 07	13	9	4	13	4	1	7	2	15	7
Metal Dep 08	5	3	9	4	2	7	9	3	1	5
Metal Dep 09	4	7	5	4	3	13	1	9	5	9

Table 8. Result for Mechanical Qualification Process for all Metal Deposition Machines

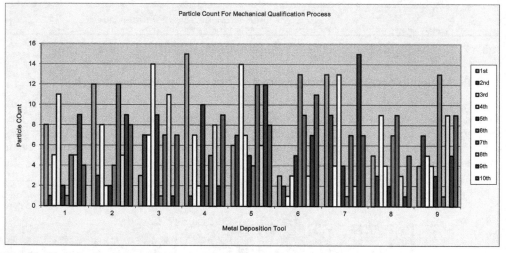

Fig. 19. Particle Count For Mechanical Qualification Process

4.4.3 Relationship between shutdown event and particle generation in gas line

Particle sensor is mounted at the gas line as shown in Figure 14 in section 3.7, to monitor particle existence in gas line. Since the gas lines are equipped with in-line gas filters, the particles are trapped in the filter. Zero reading was captured from the particle sensor. From this study, conclusion can be made that particle is not caused by the gas lines.

4.4.4 Relationship between shutdown event and sputter target life

Result for weekly shutdown event versus target life in Kilowatt per Hour (KW/H) is measured (refer Figure 20). By using Minitab, normality test was done for the target life as shown in Figure 21. The result shows the data is normal and valid to be studied.

Correlations test between shutdown and average sputter target life have been done. The Pearson correlation between shutdown and move is 0.981, which means strong correlations between both variables.

Regression analysis was made between shutdown event and sputter target life. Result for R^2 value of 96.3% indicates strong relationship between shutdown event and sputter target life. Conclusion can be made that shutdown is highly influenced by sputter target life.

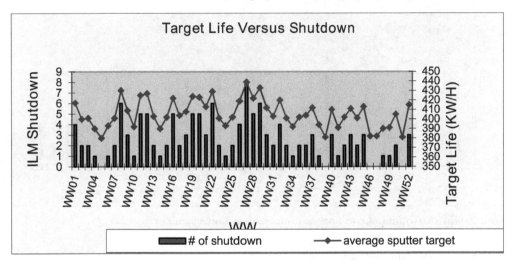

Fig. 20. Average Target Life (KW/H) versus ILM Shutdown (weekly)

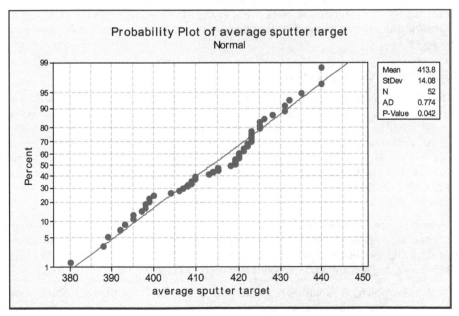

Fig. 21. Normality Test for Sputter Target Life

4.5 Environment

Result for the load lock environment is shown in Figure 22. Zero lots were captured with particle count more than three in the test. Since the specification is less than three counts, it can be concluded that load lock environment is free from particle.

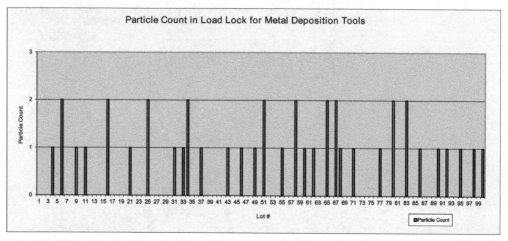

Fig. 22. Particle Count in Load Lock at Metal Deposition Machines

5. Analysis of true causes

From the verification process of potential true causes, wafer transfer in load lock and sputter target life has the most significant relationship to shutdown event. Since the most significant root cause is the sputter target life, this project will only focus on the improvement of sputter target life.

5.1 Sputter target life improvement

From the data of shutdown event versus sputter target life, observation can be made that the rate of machine shutdown is increasing by the increment of sputter target life. Even though the specification for the maximum sputter target life is 450KW/H, however in the study conducted, it shows that the chances of machine shutdown is higher if the sputter target life reach more than 410 KW/H. Zero shutdown per week was observed for sputter target life less than 390 KW/H, ten cases of one shutdown event per week were observed for sputter target life between 390 KW/H- 400 KW/H and 15 cases of two shutdown events per week were observed for sputter target life between 401 KW/H- 410KW/H.

In this study, three machines were selected to be improved by reducing the life of sputter target. Sputter target life is limited to 400 KW/H only, before replace with another new sputter target. Observation of the shutdown event versus the new sputter target life was monitored for three months, and the result is shown in Table 9.

From the data obtained, shutdown event was significantly improved. Three cases of shutdown event were observed within 12 weeks, however the cases were not related to metal deposition machine, but more to incoming particles from previous process.

Since the changes showed significant improvement in the shutdown event related to the particle issues, the new sputter target life reduction from 450 KW/H to 400 KW/H is introduced to the other six machines. Machine specification was updated with this new improvement and Preventive Maintenance job has also been revised to change the sputter target life when it reaches ~ 400 KW/H.

WW	Tool 1	2	3	Total	# of shutdown
WW01	2,875	2,790	2,698	8.363	0
WW02	2,650	2,923	2,596	8.169	0
WW03	2,748	2,866	2,777	8.391	1
WW04	2,931	3,257	2,956	9.144	0
WW05	3,096	2,385	2,343	7.824	0
WW06	2,386	3,397	2,248	8.031	0
WW07	2,847	2,711	3,464	9.022	1
WW08	3,186	3,290	3,355	9.831	0
WW09	3,727	3,513	3,628	10.868	0
WW10	3,726	3,053	3,972	10.751	0
WW11	3,527	3,331	3,319	10.177	0
WW12	3,459	3,175	3,736	10.370	1

Table 9. Shutdown versus Move base on new Sputter Target Life

6. Conclusion

Significant improvement can be seen in terms of In-Line Monitoring (ILM) shutdown event after the improvement of new sputter target life. Even though tool shutdown event still appear, it is mainly related to incoming process factors and not due to metal deposition machine. Therefore it is crucial to change the sputter target life when it reaches ~ 400 KW/H.

7. References

Alfred, M. 2001. Reduction of Particulate Contamination in Wafer Processing, pp. 220-225. Santa Clara, California: Applied Materials INC.

Bates, S.P., 2000. Silicon Wafer Processing, Applied Material Summer, US. http://www.mi.e-atech.edu/jonathan.colton/me4210/waferproc.pdf. Accessed on 24 March 2010.

Benjamin, Y.H., Ho-Ahn, L.K. 1987. Particle Technology Laboratory. Mechanical Engineering Department, University of Minnesota, Minneapolis, England.

Bordon, P. 1990. The Nature of Particle Generation in Vacuum Process Machines. IEEE Transactions on Semiconductor Manufacturing 3:189-194.

Chen, D., Seidel, T., Belinski, S. and Hackwood, S. 1989. Dynamic Particulate Characterization of a Vacuum Load- lock System. J. Vac, Science Technology 7:3105-3111.

Chowdhury, N.M. 1997. Designing a Bulk Specialty Gas System for High-Purity Applications. Proceedings of the Institute of Environmental Sciences, pp. 65-72.

Cooper, D.W. 1986. Aerosol Science Technology, pp. 25-34. New York: McGraw- Hill.

Faraday, M. 1857. The Bakerian lecture: experimental relations of gold (and other metals) to light, Philosophical Transactions of the Royal Society of London 147:145-181.

Grove, W.R. 1852. Electro-chemical polarity of gases. Philosophical Transactions of the Royal Society of London 142:87-101.

Hamberg, O.1985. Process Annual technical Meeting of Institute of Environment Sciences, Larrabee, G.B.

Hart, J. and Paterson, A. 1994. Evaluating the Particle and Outgassing Performance of High-Purity. Electronic-Grade Specialty Gas Cylinders Microcontamination 12:63-67.

Hart, J., McDermott, W., Holmer, A. and Natwora J. 1995. Particle Measurement in Specialty Gases, Solid State Technology 38:111-116.

Martin, R.W. 1989. Defect Density Measurement, in Proc. 9th International Symposium Contamination, Los Angeles.

McDermott, W.T. 1997. A Gas Diluter for Measuring Nanometer-Size Particles in Oxygen or Hydrogen. Proceedings of the Institute of Environmental Sciences and Annual Technical Meeting, pp. 26-33

Wu, J.J, Cooper, D.W. and Miller, R.J. 1989. An aerosol model of particle generation during pressure reduction, Journal of Vacuum Science & Technology A: Vacuum, Surfaces, and Films 8:1961-1968.

3

The Fundamentals of Global Outsourcing for Manufacturers

Aslı Aksoy and Nursel Öztürk

Uludag University Department of Industrial Engineering, Bursa
Turkey

1. Introduction

Today, international competition is growing rapidly, and enterprises must always remain ahead of the competition to ensure their survival. Therefore, firms must keep pace with dynamic conditions and rapid changes, be innovative, and adapt to new systems, techniques and technologies. In a competitive market environment, customers are becoming more conscious and tend to demand a particular number of customised products at a particular speed. Furthermore, fluctuations in national economies and in the global economy create significant risks. Because of all of these factors in today's competitive environment, firms have begun to make radical changes in their management and production structures. They must also reduce costs to maintain their current position in the market.

Manufacturers must be the forerunners in the competitive race in today's global markets. Today's enterprises are facing fierce competition, which is forcing them to seriously consider new applications that they can use to improve quality and to reduce cost and lead time. Manufacturers must keep pace with the dynamic requirements of the market and be receptive to reform.

Because of the intense global competition among manufacturers, the supply chain must be able to respond quickly to changes, and customer–supplier relationship management is becoming increasingly important. In recent years, very few manufacturers have owned all of the activities along the supply chain. The ability to make rapid and accurate decisions within the supplier network improves the competitive advantage of manufacturers.

Additionally, due to the intense global competition that exists today, firms should be reevaluate and redirect missing resources. Outsourcing plays a key role for enterprises because the cost of raw materials constitutes a significant part of the cost of the final product. Choosing the right supplier reduces purchasing costs and enhances the competitive advantage of firms. As organisations become more dependent on suppliers, the direct and indirect consequences of poor decision-making become more severe. Decisions about purchasing strategies and operations are the primary determinants of profitability. The globalisation of trade and the Internet have enlarged purchaser choice sets. Changing customer preferences require broader, more systematic and faster outsourcing decision making.

An enterprise may produce a specific product itself or may outsource that specific product to achieve a production cost advantage. Global outsourcing can be defined as the forwarding of specific business to a global supplier. Global outsourcing enhances the competencies of firms while also making firm structures more flexible.

In today's global markets, firms must use new methods to sustain their strength and compete. In recent years, under the influence of this intense competition, global outsourcing has become popular for firms. Firms are widely using global outsourcing to adapt to rapid changes, to reduce the effect of fluctuations, and to take advantage of know-how and current technologies. Global outsourcing allows firms to develop their core competencies and expand their flexibility.

This study reviews the literature on global outsourcing. There has been a great deal of research conducted on global outsourcing within information technology (IT) and service systems. To the best of our knowledge, there have not been many studies on the global outsourcing of manufacturing or production systems. Therefore, this study focused on the global outsourcing of these systems. This paper provides a basic definition of global outsourcing and analyses global outsourcing as either an opportunity or a threat. Furthermore, this study introduces the differences between local and global outsourcing. The methods used to make global outsourcing decisions and the decision criteria used in global outsourcing are also presented in the study.

2. Global outsourcing

Outsourcing is one of the responsibilities of purchasing departments and plays a critical role in an organisation's survival and growth. Materials sourced from outside rather than produced by in-house facilities will influence service quality and profitability (Zeng, 2000). Despite the ongoing debate over the benefits and risks of outsourcing for businesses, outsourcing has become a common approach that purchasing managers cannot ignore. Indeed, outsourcing has exhibited dramatic growth in recent years.

Since the 1980s, the opinions of purchasing managers and management scholars regarding optimal firm sourcing strategies have changed significantly in two respects. First, firms have replaced vertical integration with increased outsourcing based on the conviction that lean, flexible enterprises that focus on their core competencies perform better (Quinn & Hilmer, 1994). Second, in the era of globalisation of the 1990s, enterprises were advised to use the principles of "global outsourcing" to pick the best global suppliers and thereby to improve their competitiveness (Monczka & Trent, 1991; Quinn & Hilmer, 1994). Implementing both or either of these strategies has important consequences for the structure and performance of multinational corporations.

Given the rapidly shifting contours of the global economy, companies need to be able to anticipate changes in the economics and geography of outsourcing. Forward-thinking companies are making their value chains more elastic and their organisations more flexible. Furthermore, with the decline of the vertically integrated business model, outsourcing is evolving into a strategic process used to organize and fine-tune the value chain.

Supplier selection and evaluation play an important role in reducing the cost and time to market while improving product quality. Supplier selection can significantly affect manufacturing costs and production lead time. Although several techniques and models have been used to select and evaluate suppliers, each technique or model has its own strengths and limitations in different situations. Therefore, it is necessary to further improve the performance and effectiveness of supplier selection and evaluation in manufacturing in different contexts.

According to Boer et al. (2001), the purchasing function and purchasing decisions are becoming increasingly important. As organisations become more dependent on suppliers, the direct and indirect consequences of poor decision making are becoming more severe. In addition, several developments have further complicated purchasing decision making. The globalisation of trade and the Internet have expanded the choice sets of purchasers. Changing customer preferences requires broader and more rapid supplier selection.

Outsourcing can be defined as the provision of services by an outside company when those services were previously provided by the home company. In other words, outsourcing involves focusing on a firm's core competencies while allowing services that require other competencies to be provided by other expert enterprises. Outsourcing is a strategic decision in which the buying firm attempts to establish a long-term business relationship with its suppliers (Zeng, 2000).

It is not always easy to generate precise rules for the supplier selection process, but certain elements of the process remain constant. These elements may be identified based on intuition, experience, common sense, or inexplicable rules. Supplier ratings, for example, are usually generated via subjective criteria, based on personal experience and beliefs, based on the available information, and/or sometimes using techniques and algorithms intended to support the decision-making process (Albino & Garavelli, 1998). The key to enhancing the quality of decision making in the supplier selection process is to employ the powerful computer-related concepts, tools and techniques that have become available in recent years (Wei et al., 1997).

In today's competitive global markets, consumers look for the highest quality products at the lowest prices, regardless of where they are produced. This trend is continuously increasing the significance of global markets and forcing enterprises to enter global markets. Furthermore, increasing pressure from foreign competitors in domestic markets is forcing companies to analyse the available alternatives as they seek to remain competitive.

Monczka & Trent (1991) defined global outsourcing as the integration and coordination of procurement requirements across worldwide business units. As such, outsourcing might involve objects, processes, technologies and suppliers. Kotabe (1998) defined global outsourcing as the purchase of finished products or works-in-process from global suppliers. Under this definition, firms may purchase not only products themselves but also the services required to make these products marketable.

Narasimhan et al. (2006) reported that the strategic objectives of global outsourcing are different from those of traditional purchasing. Whereas traditional purchasing focuses on minimizing procurement costs, strategic global outsourcing considers quality, delivery, responsiveness and innovativeness in addition to costs. Sourcing strategies should be incorporated into the operating strategies of buying firms to support or even improve their competitive advantage (Tam et al., 2007). Internal or global outsourcing plays an important role in firm competitiveness and growth (Zeng, 2000).

Flexibility appears to be an important driver of global outsourcing strategy. Firms need to react more quickly to customer requirements, and global outsourcing is seen as a way to accomplish this. Global outsourcing may also be perceived as a way to reduce firm risk by sharing it with suppliers and simultaneously acquire the positive attributes of those suppliers (Kremic et al., 2006). The ultimate objective of global outsourcing strategy is for the firm to exploit both its own and suppliers' competitive advantage and to utilise the comparative location advantages of various countries in global competition (Kotabe & Murray, 2004).

The importance of global outsourcing has increased dramatically. Although firms may outsource for cost-related reasons, there are no guarantees that expected savings will be achieved (Kremic et al., 2006).

Global outsourcing strategy requires close coordination between the research and development, manufacturing, and marketing activities of a firm. Conflicts will most likely exist between the differing objectives of these divisions. For instance, excessive product modification and development intended to satisfy a set of ever-changing customer needs will negatively affect manufacturing efficiency and increase costs. Similarly, excessive product standardization intended to lower manufacturing costs will likely yield lower customer satisfaction levels (Kotabe & Murray, 2004). Therefore, effective global outsourcing requires firms to develop a balance between effective manufacturing and flexible marketing.

Global outsourcing is an expected response to competition. However, the choice of where to obtain goods and services is not an obvious decision. Rather, it is subject to continual reevaluation (Carter et al., 2008). Outsourcing strategy is an essential part of the value chain for corporate activities. Outsourcing strategy both affects and is affected by the other aspects of the supply chain (Kotabe et al., 2008).

The degree of internationalization of production and sourcing is negatively related to the size of the focal country. According to Mol et al. (2005) and Buckley & Pearce (1979) when working with a sample of 156 Japanese, French, Swiss, and "Benelux" companies, found the ratios of global outsourcing to final markets to be 2.4%, 8.0%, 91.6%, and 70.7%, respectively.

As Levy (2005) noted, global outsourcing is highly related to efforts to increase the organizational and technological capacity of firms. Mol et al. (2005) described global outsourcing as balancing international production cost advantage and domestic transaction cost advantage rather than characterising it as a performance-enhancing tool. The major operational problems in global outsourcing, as described by Kotabe et al. (2008), are logistics, inventory management, distance, nationalism, and a lack of working knowledge about foreign business practices.

Global outsourcing has become a popular subject of study in both managerial practice and the academic literature. Conflicting results have been presented in the relevant studies. The global outsourcing strategy literature offers arguments both for and against global outsourcing strategy (Kotabe et al., 2008).

According to Gottfredson et al. (2005), a recent survey of large and medium-sized companies indicates that 82% of large firms in Europe, Asia, and North America have outsourcing arrangements of some kind and that 51% use global outsourcers. However, nearly 50% say that the results achieved by their outsourcing programs have fallen short of expectations. What is more, only 10% are highly satisfied with the decreases in costs that they have achieved, and a mere 6% are highly satisfied with the results of their global outsourcing efforts overall. Mol et al. (2005) stated that global outsourcing can help a firm to enhance its competitive advantage in other markets or to improve its legitimacy. However, multinational supply chains are facing significant managerial problems related to international relations.

According to Kremic et al. (2006), the expected benefits of outsourcing may include providing the same or a better service at a lower overall cost, increased flexibility and/or quality, access to the latest technology and the best talent, and the ability to refocus scarce resources on core functions.

A lack of common methodology is believed to cause some outsourcing failures. Lonsdale (1999) also supported this thinking, suggesting that global outsourcing failures are not due to inherent problems with outsourcing but rather stem from a lack of guiding methodology for managers. Kremic et al. (2006) indicated that global outsourcing has potential pitfalls for strategic reasons. Gillett (1994) noted that enterprises may lose their core competences if they are not careful. If firms outsource the wrong functions, they may develop gaps in their learning or knowledge base that may hinder their ability to capitalise on future opportunities (Kremic et al., 2006). Literature also indicated that in industries with complex technologies and systems, internal synergy may decrease when some functions are outsourced. This could result in lower productivity or efficiency levels for the remaining functions (Quinn & Hilmer, 1994).

Kremic et al. (2006) discussed factors that may impact global outsourcing decisions. These factors are shown below:

- *Core competences*: "Core competences" can be described as a strategic factor that firms use to sustain competitive advantage. Quinn (1999) suggested that there are "core activities" that one firm will perform better than any other firm. In general, a function that is more core to an organization is less likely to be globally outsourced.
- *Critical knowledge:* Some data or knowledge must be under the control of the firm. In general, if a function provides critical knowledge, it is less likely to be globally outsourced.
- *Impact on quality:* The quality of the firm's services establishes its reputation and can create demand. If a firm is currently recognised in the industry for providing a high level of quality in a particular area, then global outsourcing in that area can harm quality. Quality is a relevant factor and can have either a positive or a negative influence on global outsourcing decisions. (Anderson, 1997).
- *Flexibility:* Flexibility includes demand flexibility, process flexibility and resource flexibility. Antonucci et al. (1998) noted that long contracts outsourced into a limited market have sometimes decreased flexibility. However, large enterprises may improve their flexibility via global outsourcing. In the literature, global outsourcing is used as a strategic driver to increase flexibility.
- *Cost:* In the literature, cost is the main reason for global outsourcing decisions. If the firm prefers to outsource a function for cost reasons, then it can be assumed that the current expenditures associated with that function are higher than the expected cost of purchasing the service. However, whether savings will actually accrue from global outsourcing is extremely uncertain. Sometimes, the reported cost savings may not be as high as was expected.

- *Characteristics of the functions outsourced or kept in-house:* In general, the more complex a function the less of a candidate it is for global outsourcing.
- *Integration:* Integration refers to the degree to which function is linked to other functions and systems within the enterprise. The more integrated the function, the more interactions and communication channels there are to maintain and monitor. Therefore, a function that is highly integrated is less of a candidate for global outsourcing.

Firms establish and execute global outsourcing plans in an effort to match competitors' attempts at outsourcing; improve non-competitive cost structures; focus on core competencies; reduce capital investment and overall fixed costs; achieve cost-competitive growth within their supply base for goods, services and technologies in the value chain; and establish future sales footprints in low-cost countries by outsourcing basic goods or business processes (Carter et al., 2008). An effective global sourcing strategy requires continual efforts to streamline manufacturing without sacrificing marketing flexibility (Kotabe & Murray, 2004).

According to the literature, firms prefer global outsourcing for the following reasons:

- *Strategic focus / reduction of assets:* Through global outsourcing activities, an enterprise can reduce its level of asset investment in manufacturing and related areas. Furthermore, global outsourcing can help management teams to redirect their attention to core competencies rather than focusing on maintaining a wide range of competencies (Kotabe et al., 2008).
- *Supplementary power / lower production costs:* Global suppliers are highly specialized in their own business, which lowers both their production costs for those of the firms that are outsourcing their business to them. Therefore, global outsourcing can decrease overall costs if firms globally outsource non-core activities (Quinn, 1999)
- *Strategic flexibility:* Global outsourcing can enhance a firm's strategic flexibility (Harris et al., 1998). If a firm is faced with a crisis in an external environment, it can simply change the volume of globally outsourced products it purchases. If the same product is outsourced to another firm within the home country, the firm will need to pay high reconstructing costs and may not respond quickly to the external environment.
- *Relationship:* Certain relationships with global suppliers can deliver competitive advantage for firms (Kotabe et al., 2008). Misunderstandings between buyer and suppliers may decrease a firm's level of performance (Carter et al., 2008).

Kotabe et al. (2008) suggested that an inverted-U shaped relationship exists between profitability and the degree of outsourcing. On the inverted-U shaped curve, there is an optimal degree of outsourcing for a firm. If the firm moves' away from this optimal point, profitability decreases dramatically.

In global outsourcing strategy, there are also some disadvantages of increasing total product cost. Unfortunately, through global outsourcing, the cost of transportation, communication and information-sharing may increase. Domestic purchasing strategies require only short lead times because they reduce communication and transportation time requirements. The literature suggests that this may be the key reason why some enterprises do not prefer global outsourcing (Dana et al., 2007).

The literature suggests some disadvantages of global outsourcing. These disadvantages can be seen below:

- *The scope of the functions:* If there are important interfaces between activities, decoupling them into separate activities performed by separate suppliers will generate less than optimal results and potential integration problems (Kotabe et al., 2008).
- *Competition loss:* Firms that engage in excessive outsourcing are essentially hollowing out their competitive base (Kotabe, 1998). Furthermore, an enterprise may lose negotiating power with its suppliers because the capabilities of the latter will increase relative to those of the former (Kotabe et al., 2008).
- *Opportunistic behaviour:* Global suppliers may behave opportunistically. Opportunistic behaviour allows a supplier to extract more rents from the relationship than it would normally do, for example, by supplying products of a lower quality than was previously agreed upon or by withholding information regarding changes in production costs (Kotabe et al., 2008).
- *Limited learning and innovation:* Suppliers may capture the critical knowledge by performing the activity. This situation is always a problem between buyer and supplier because both try to obtain all the individual benefits. Appropriation of innovations and rents is always a problem in such a complex buyer–supplier relationships (Nooteboom, 1999)
- *Negative impact of exchange rates:* Higher procurement costs can be seen by the negative impact of fluctuating exchange rates. During the Asian financial crisis, many foreign firms operating in Asian countries learned an invaluable lesson on the negative impact of fluctuating currency exchange rates on their procurement costs and profitability (Kotabe et al., 2008).

According to Kremic et al. (2006), the global outsourcing literature has referenced the following risks of global outsourcing: the potential for both unrealized savings and increased costs, employee morale problems, over-dependence on suppliers, lost corporate knowledge and future opportunities, and under-satisfied customers. Additionally, global outsourcing may fail because the requirements of the relationship are inadequately defined because of a poor contract, a lack of guidance regarding planning or managing outsourcing initiatives, or poor supplier relations. Dana et al. (2007) cited lower production costs as the key advantage of a global outsourcing strategy, with poor control of quality being the main disadvantage.

Lowe et al. (2002) addressed two risks of global outsourcing: fluctuations in exchange rates and relative rates of inflation in different countries. The impact of fluctuations in exchange rates can be analyzed in different ways, and these disparate analyses can yield different results. Brush et al. (1999) stated that many enterprises do not discuss exchange rates as a key factor in global outsourcing. Kouvelis (1999) stated that because of the high cost of switching global suppliers, purchasing managers do not switch suppliers until the effect of exchange rate fluctuation is extremely high. Vidal & Goetschalckx (2000) indicated that the impact of exchange rate fluctuation on overall cost is high.

Under competitive pressure, many U.S. multinational companies globally outsource components and finished products to countries such as China, South Korea, Taiwan, Singapore, Hong Kong, and Mexico. Those countries are also known as low-cost countries (Kotabe & Murray, 2004). Firms in the US and the EU make different choices when selecting global outsourcing locations. In the US, 23% of enterprises prefer China, 14% prefer India, 10% prefer Mexico, 9% prefer Argentina and 8% prefer Brazil. In the EU, 19%

prefer China, 14% prefer the Czech Republic, 12% prefer Poland and 10% prefer Hungary (Timmermans, 2005). The preferences of US and EU firms indicate what is known as "low-cost country sourcing" in the literature. Low-cost country sourcing entails the sourcing of services or functions from low-cost countries with lower labour and material costs. In recent years, low-cost country sourcing has created opportunities for purchasing managers (Carter et al., 2008).

Sourcing from global suppliers can be risky, especially when the projected quality of the outsourced products is unknown. Motwani et al. (1999) noted that as the low-cost countries develop, the quality of the products produced in those countries will likely increase. As a result, firms that choose to forge relationships in these low-cost countries now through sourcing and purchasing may have an edge in these markets in the future. Although they may encounter challenges at first, the advantages that they enjoy in the future could outweigh these problems. This may be especially true for firms that aim to be truly global. Although the main factor driving global outsourcing is lower costs, experienced purchasing managers consider many factors simultaneously in making the decision to outsource internationally. According to the relevant literature, lower labour cost is not the key factor for many US enterprises that engage in global/domestic outsourcing (Sarkis & Talluri, 2002).

3. Outsourcing methods in literature

Outsourcing has been widely discussed in the literature. There are several papers on supplier selection and global outsourcing in information technology (IT) and for service systems. To the best of our knowledge, few studies have been conducted on global outsourcing in manufacturing or production systems. This section is divided into two subsections: one that addresses general supplier selection methods in the literature and another that addresses global outsourcing methods in the literature.

3.1 General supplier selection methods in the literature

There are several methods of general supplier selection presented in the literature. Categorical methods are qualitative models. Based on the buyer's experience and historical data, suppliers are evaluated using a particular set of criteria. The evaluations involve categorizing the supplier's performance as 'positive', 'neutral' or 'negative' with reference to a series of criteria (Boer et al., 2001). After a supplier has been rated for all the criteria, the buyer provides an overall rating, allowing the suppliers to be sorted into three categories.

Data Envelopment Analysis (DEA) is concerned with the efficiency of decision making. The DEA method helps buyers to classify suppliers into two categories: efficient suppliers and inefficient suppliers. Liu et al. (2000) used DEA in the supplier selection process. They evaluated the overall performance of suppliers using DEA. Saen (2007) used IDEA (Imprecise Data Envelopment Analysis) to select the best suppliers based on both cardinal and ordinal data. Wu et al. (2009) proposed an augmented DEA approach to supplier selection. Songhori et al. (2011) presented a structured framework for helping decision makers to select the best suppliers for their firm using DEA.

Cluster Analysis (CA) is a class of statistical techniques that can be used with data that exhibit "natural" groupings (Boer et al., 2001).

Case-Based Reasoning systems (CBR) combine a cognitive model describing how people use and reason from past experience with a technology for finding and presenting experience (Choy et al., 2003-a). Choy et al. (2002-b) enhanced a CBR-based supplier selection tool by combining the Supplier Management Network (SMN) and the Supplier Selection Workflow (SSW). Choy et al. (2005) used CBR to select suppliers in a new product development process.

In linear weighting, the criteria are weighted, and the criterion with the largest weight has the greatest importance. The score for a particular supplier is based on the criteria and their different levels of importance, and some criteria have a high degree of precision. Ghodsypour & O'Brien (1998) integrated the Analytic Hierarchy Process (AHP) and linear programming to consider both tangible and intangible factors in choosing the best suppliers and the optimum order quantities. Lee et al. (2001) used only the AHP for supplier selection. They determined the supplier selection criteria based on purchasing strategy and criterion weights using the AHP. Liu & Hai (2005) used DEA to determine the supplier selection criteria. They then interviewed 60 administrators to determine the priority level of the criteria and used the AHP to select suppliers. Ting & Cho (2008) presented a two-step decision-making procedure. They used the AHP to select a set of candidate suppliers for a firm and then used a Multi-Objective Linear Programming (MOLP) model to determine the optimal allocation of order quantities to those suppliers. Boer et al. (1998) used the ELECTRE 1 technique to evaluate the five supplier candidates. Xia & Wu (2007) used an integrated approach to the AHP, which was improved using rough set theory and multi-objective mixed integer programming to simultaneously determine the number of suppliers to employ and the order quantities to be allocated to these suppliers in the case of multiple sourcing and multiple products. Multiple criteria and supplier capacity constraints were both taken into account. Wang et al. (2004) used an integrated AHP and preemptive goal programming (PGP)-based multi-criteria decision-making process to analyze both the qualitative and quantitative factors guiding supplier selection. Liu and Hai (2005) compared the use of the Voting Analytic Hierarchy Process (VAHP) and the use of the AHP for supplier selection. Chan & Kumar (2007) identified some of the important decision criteria, including risk factors in developing an efficient system of global supplier selection. They used the Fuzzy Extended Analytic Hierarchy Process (FEAHP) to select suppliers. Chan & Chan (2010) used an AHP-based model to solve the supplier evaluation and selection problem for the fashion industry. Kumar & Roy (2011) proposed the use of a rule-based model with the AHP to aid decision makers in supplier evaluation and selection.

Total Cost of Ownership models (TCO) include all costs related to the supplier selection process that are incurred during a purchased item's life cycle. Degraeve & Roodhooft (1999) evaluated suppliers based on quality, price and delivery performance using TCO. They emphasised that uncertainty related to demand, delivery, quality and price must be reflected in the decision problem. Ramanathan (2007) proposed the integrated DEA-TCO-AHP model for supplier selection.

According to Boer et al. (2001), Mathematical Programming models (MP) allows the decision maker to formulate the decision problem in terms of a mathematical objective function that must subsequently be maximized and minimized by varying the values of the variables in the objective function. MP models are more objective than rating models because they force the decision maker to explicitly state the objective function, but MP models often only consider more quantitative criteria. Karpak et al. (1999) developed a supplier selection tool that minimizing costs and maximizing quality reliability. Ghodsypour & O'Brien (1998) integrated the AHP and Linear Programming (LP) models. Their model presented a systematic approach that took into account both qualitative and quantitative criteria. They also developed sensitivity algorithms for different scenarios. Ghodsypour & O'Brien (2001) used mixed integer programming, taking into account the total cost of logistics. Degraeve & Roodhooft (2000) computed the purchasing cost associated with different purchasing strategies using MP. Barla (2003) reduced the number of suppliers from 58 to 10 using the multi-criteria selection method. Hong et al. (2005) decomposed the supplier selection process into two steps. They used cluster analysis to pre-select suppliers and then used MP to select the most appropriate supplier. Yang et al. (2007) studied a supplier selection problem in which a buyer facing random demand must decide the quantity of products it will order from a set of suppliers with different yields and prices. They provided the mathematical formulation for the buyer's profit maximization problem and proposed a solution method based on combining the active set method and the Newton search procedure. Kheljani et al. (2007) considered the issue of coordination between one buyer and multiple potential suppliers in the supplier selection process. In contrast, in the objective function in the model, the total cost of the supply chain is minimized in addition to the buyer's cost. The total cost of the supply chain includes both types of costs. The model was solved using mixed-integer nonlinear programming. Liao & Rittscher (2007) developed a multi-objective programming model, integrating supplier selection to procure lot sizing and carrier selection decisions for a single purchasing item over multiple planning periods during which the demand quantities are known but inconstant. Rajan et al. (2010) proposed a supplier selection model for use in a multiproduct, multi-vendor environment based on an integer linear programming model.

Artificial intelligence (AI)-based systems are computer-aided systems that can be trained using data on purchasing experience or historical data. The available types of AI-based supplier selection applications include Neural Networks (NN) and Expert Systems (ES). One of the important advantages of the NN method is that the method does not require the formulation of the decision-making process. As a result, NNs can cope better with complexity and uncertainty than traditional methods can; these systems are designed to be more similar to human judgment in their functioning. The system user must provide the NN with the properties of the current case. The NN provides information to the user based on what it has learned from the historical data. Albino & Garavelli (1998) further developed the neural network-based decision support system for subcontractor ratings in construction firms. The system includes a back-propagation algorithm. The constructed network is trained using examples so that the system does not require decision-making rules. Vokurka et al. (1996) and Wei et al. (1997) developed an expert system for supporting the supplier selection process. Chen et al. (2006) used linguistic values to assess the ratings and weights of various supplier selection factors. These linguistic ratings were expressed using trapezoidal or triangular fuzzy numbers. Then, they proposed the use of a hierarchy Multiple Criteria Decision-Making (MCDM) model based on fuzzy-sets theory to address supplier selection problems in the supply chain system.

Wang & Che (2007) presented an integrated assessment model for manufacturers to use to solve complex product configuration change problems efficiently and effectively. The model made it possible to determine what fundamental supplier combination would best minimize the cost–quality score if and when proposed by the customer and/or engineer. The researchers combined fuzzy theory, T transformation technology, and genetic algorithms. Liao & Rittscher (2007) studied the supplier selection problem under stochastic demand conditions. Stochastic supplier selection is determined by simultaneously considering the total cost, the quality rejection rate, the late delivery rate and the flexibility rate while also taking into account constraints on demand satisfaction and capacity. The researchers used GA to solve the problem. Wang (2008) developed a decision-making procedure that could be used for supplier selection when product part modifications were necessary. The aim of the research was to determine acceptable near-optimal solutions within a short period of time using a solution-finding model based on Genetic Algorithms (GA). Aksoy & Öztürk (2011) presented a neural network-based supplier selection and supplier performance evaluation system for use in a *just-in-time* (JIT) production environment. Chang et al. (2011) proposed the use of a fuzzy decision-making method to identify evaluation factors that could be used for supplier selection. Jiang & Chan (2011) proposed a method of using a fuzzy set theory with twenty criteria to evaluate and select suppliers.

3.2 Global outsourcing methods in the literature

Canel & Khumawala (1996) proposed a 0-1 mixed integer programming formulation model for international facilities location problems. They determined the location of the international facility and the capacity of that facility. The objective of the model is to maximize the after-tax profit. The proposed model includes different costs, including investment cost, fixed costs, transportation costs, shortage costs and holding costs. The researchers developed two different mathematical models: one for a capacitated case and the other for an uncapacitated case. They used demand and price as the deterministic parameters. Their research could be extended by relaxing the assumptions of deterministic demand, prices, costs, etc. within the problem and treating those factors as stochastic parameters. Huchzermeier & Cohen (1996) developed a stochastic dynamic programming formulation for evaluating global manufacturing strategy options while taking switching costs into account in a stochastic exchange-rate environment. The objective of the model is to maximize after-tax profits. The model includes taxes, fixed and variable costs, capacity and exchange rates. The decision variable in the model is production quantities. The researchers developed different scenarios for different exchange rates. Each model has its own solution. However, the model does not include qualitative parameters. Canel & Khumawala (1997) presented an efficient branch-and-bound procedure for solving uncapacitated, multi-period international facilities location problems. The branch-and-bound problems can be solved using LINDO. The parameters of the model are assumed to be deterministic. Dasu & Torre (1997) presented a model for planning a global supply network for a multinational yarn manufacturer. The objective of the model is to maximize the overall profits of the global supply chain network. The model includes tariffs, exchange rates and transportation costs. The proposed model is non-linear, but the authors make some assumptions in the model to give it a linear structure.

Kouvelis & Gutierrez (1997) solved the newsvendor problem in the textile industry for "style goods". The proposed model determines the production quantities while minimizing shortages and holding costs for a multiple-location manufacturer in a multiple-location market. Shortage costs in this context include the costs associated with lost sales, and holding costs includes the cost associated with excess inventory left over after the selling season. The model includes transportation costs, exchange rate uncertainty and stochastic demand uncertainty, but the model does not include global cost factors such as taxes and tariffs. The study evaluates alternative plans for supply chain design and centralized and decentralized production decision-making mechanisms. The researchers also stated that centralized production decision-making is superior to decentralized production decision-making but that application and control problems are associated with centralized coordination. The proposed model can be easily implemented by purchasing managers. The researchers noted that the production decision-making process can be affected by the uncertainty of global markets and that models with stochastic parameters (such as models for analysing political risk and exchange rate fluctuations) can be used in future research. Munson & Rosenblatt (1997) described local content rules and developed models for selecting global suppliers while satisfying local content provisions. The parameters of the model are deterministic, and the penalty for breaking local content rules is very high. The researchers used the mixed integer programming method to solve the model. The decision variables for the model are the selection of the global supplier and the allocation of orders among the selected suppliers. The objective of the model is to minimize purchasing, production, transportation and fixed costs. The model considers only costs and local content rules. The model does not take into account quantitative parameters or exchange rates.

Coman & Ronen (2000), formulated the global outsourcing problem as a linear programming (LP) problem, identified an analytical solution, and compared that solution with the solutions obtained using the standard cost accounting model and the theory of constraints. The decision variable for the model is the production quantity in terms of preference to manufacture versus preference to outsource. The solution attained indicated that linear programming yielded better results than the two other methods (standard cost accounting and the theory of constraints).

Canel & Khumawala (2001) solved an international facilities location problem using the heuristic method. They developed 12 heuristic methods, but their models do not include quantitative parameters or exchange rates. Vidal & Goetschalckx (2001) presented a model for optimizing global supply that maximizes after-tax profits for a multinational corporation. The model includes transfer prices and the allocation of transportation costs as explicit decision variables. Transfer prices and flows between multinational facilities are calculated in the model. The model does not address the supplier selection problem. The model entails a non-convex optimization problem with a linear objective function, a set of linear constraints, and a set of bilinear constraints. Because the resulting problem is NP hard, the researchers developed a heuristic successive linear programming solution procedure.

Canel & Das (2002) proposed the use of a 0-1 mixed integer programming model to determine for particular time periods which countries a firm should choose as the locations of its global manufacturing facilities. The model was also used to determine the quantity to be produced at each global manufacturing facility and the quantities to be shipped from the global facilities to customers. The study had two goals: to determine the location of the global manufacturing facility and to develop a mathematical model that included global marketing and manufacturing factors. The proposed model does not include quantitative parameters or take exchange rates into account. Hadjinicola & Kumar

(2002) presented a model that includes manufacturing factors together with factory location, inventory, economies of scale, product design, and postponement. In the study, different manufacturing and marketing strategies were evaluated for two different countries. The proposed model is a descriptive model; therefore, the decision variables are not entirely clear. The model can be used to evaluate different manufacturing and marketing strategies in terms of cost and profit functions. The model also includes the effect of exchange rates. Lowe et al. (2002), proposed a model that included exchange rates, using it to choose the location and capacity of global manufacturing facilities in the chemical industry. They analysed data from the year 1982, setting the production capacity, purchasing volume, capacity, fixed and variable costs, transportation costs and taxes for each production facility. To take into account the effect of exchange rates, they reviewed historical data from 22 years and constructed nine different scenarios, calculating the costs for each scenario. They developed a two-stage method of solving the problem. The first stage is a short planning period (as an example 1 year), and the second stage is an optional stage that can be used for long-term planning. However, for longer periods, the first stage can also be repeatedly used.

Teng & Jaramillo (2005) presented a model for global supplier selection in the textile industry. They weighted their criteria and sub-criteria based on expert opinions. The overall scores for the global supplier are calculated by multiplying the score and the weight of the criteria. Goh et al. (2007) presented a stochastic model of the multi-stage global supply chain network problem, incorporating a set of related risks: supply, demand, exchange, and disruption. The objective of the model is to maximize after-tax profits. The model includes demand uncertainty, exchange rates, taxes and tariffs. The researchers composed different scenarios for demand uncertainty and exchange rates and presented different clusters of stochastic parameters. Because the proposed model is a convex linear model, the authors relaxed certain parameters to make the problem linear, but theirs is a descriptive model that is not applied to a real-world scenario.

Lin et al. (2007) proposed the use of a decision model to support global decision making. The model uses two multiple-criteria decision aid techniques (the AHP and PROMETHEE II) and incorporates multiple dimensions (infrastructure, country risk, government policy, value of human capital and cost) into a sensitivity analysis. The authors used the AHP to determine the weight of the criteria and used PROMETHEE to select global suppliers based on weighted criteria. Kumar & Arbi (2008) used a simulation model to forecast lead time and total cost in a global supply environment. Based on their results, it seems that important cost savings can be attained through global outsourcing. However, lead time is an essential factor in real life. The authors stated that global outsourcing is not a viable way to meet short-term market demands but that for large seasonal orders, global outsourcing can be a significant cost-saver. Ray et al. (2008) described the cause of the outsourcing problem, formulated it as a linear programming problem, developed a corresponding function, and offered a simplified criterion for ordering products in terms of preference for manufacturing versus preference for global outsourcing. The authors used a hybrid approach, incorporating the Hurwicz criterion, the theory of constraints (TOC) and linear programming. Some weaknesses of the proposed method are that it is difficult to change the traditional cost accounting system, that it would take time to implement the approach, and that people may be reluctant to use the approach because it requires them to justify their preferences rather than simply saying yes or no. It also requires a new decision-making process.

Wang et al. (2008a) divided firm activities into core activities, core-close activities, core-distinct activities and disposable activities. They used the ELECTRE 1 method to determine which of those four activities can be globally outsourced. Wang et al. (2008b) presented a model for the global outsourcing of logistics activities. They determined the evaluation criteria, ranked the criteria using the AHP and constructed a method using PROMETHEE for global supplier selection. Feng & Wu (2009) presented several tax-saving approaches and developed a tax savings model for maximizing after-tax profit from logistics activities by global manufacturers. Using this model, logistics activities are evaluated in terms of tax savings. It has been observed that the tax saving model has dramatically increased manufacturer profits. The suppliers are selected based on tax savings, whereas any other criteria are disregarded.

Ren et al. (2009) treated global supply chains as agile supply chains and explained agility as the ability to change and adapt quickly to changing circumstances. The model facilitates supplier selection for agile supply chains. The authors determined 10 criteria and 32 sub-criteria for supplier selection. The weights of the criteria were determined based on expert opinions and used to rank the suppliers. Perron et al. (2010) presented a mathematical model for multinational enterprises to use to determine transfer prices and the flow of goods between global facilities. The model includes bilinear constraints; therefore, the authors relaxed the constraints to simplify the model. They developed a branch-and-cut algorithm and two different heuristics to solve the model. The heuristic methods can be summarised as follows:

1. *Variable Neighbour Search Method (VNS):* This method is based on the concept of systematic changes in neighbourhoods during the search. VNS explores nearby and then increasingly far neighbourhoods for the best-known solution in a probabilistic fashion. Therefore, often favourable characteristics of the best-known solution will be kept and used to obtain promising neighbouring solutions. VNS was repeatedly used with a local search routine to transition from these neighbouring solutions to local optima.
2. *Alternate Heuristic (ALT):* Given two subsets of variables, the ALT heuristic solves the problem by alternately fixing the variables of one of the subsets. The subsets of variables must be such that the model becomes linear when fixing the variables of one of the subsets. When one of these linear programs is solved, its solution becomes a set of parameters in the other one. ALT can be converging to local optima.

The objective of the model is to maximize after-tax profit given taxes, capacity, transfer prices and demand. Satisfactory results were reported when small problems were solved using heuristic methods.

4. Outsourcing decision criteria

To be competitive in the global market, firms must attain the knowledge necessary to systematically evaluate all potential suppliers and select the most suitable ones. The factors most often used in current supplier evaluation are quality, supplier certification, facilities, continuous improvement, physical distribution and channel relationships (Weber, 1991).

In the supplier selection process, it is not always easy to recognize precise rules, but there is, in general, a coherent way to solve the problem. This coherence can be rooted in intuition, experience, common sense, or inexplicable rules. Supplier rating is then a problem usually solved by subjective criteria, based on personal experiences and beliefs, on the available information and, sometimes, on techniques and algorithms supporting the decision process (Albino & Garavelli, 1998). The key to enhancing the quality of decision making in supplier selection include the powerful computer-related concepts, tools and techniques that have become available in recent years (Wei et al., 1997).

Chao et al. (1993) concluded that quality and on-time delivery are the most important attributes of purchasing performance. Ghodsypour & O'Brien (1998) agreed that cost, quality and service are the three main factors that should influence supplier selection. Brigs (1994) stated that joint development, culture, forward engineering, trust, supply chain management, quality and communication are the key requirements of supplier partnerships apart from optimum cost. Petroni & Braglia (2000) evaluated the relative performance of suppliers with multiple outputs and inputs, considering management, production facilities, technology, price, quality, and delivery compliance. Wei et al. (1997) examined factors such as supply history, product price, technological ability and transport cost.

Making sourcing decisions based on delivery speed and cost is the best way to improve performance (Tan 2001). Global outsourcing reduces the fixed investment costs of a firm in its own economic region. Today, in making global outsourcing decisions, many enterprises also consider quality, reliability, and technology when evaluating the components and products to be procured (Kotabe & Murray, 2004) rather than only considering price. In developing global outsourcing strategies, firms must consider not only manufacturing costs, the costs of various resources, and exchange rate fluctuations but also the availability of infrastructure (including transportation, communications, and energy), industrial and cultural environments, and ease of working with foreign host governments, among others (Kotabe et al., 2008).

Several factors influence global outsourcing decisions. Canel & Das (2002) outlined the factors that most commonly influence global manufacturing facility locations. Those factors are labour and other production inputs, political stability, the attitude of the host government towards foreign investment, host government tax and trade policies, proximity to major markets, access to transportation and the existence of other competitors. Choy et al. (2005) stated that good customer–supplier relationships are necessary for an organization to respond to dynamic and unpredictable changes. They considered price, delivery, quality, innovation, technology level, culture, commercial awareness, production flexibility, ease of communication and current reputation when selecting and evaluating suppliers. Teng & Jaramillo (2005) used five main criteria and 20 sub-criteria for global supplier selection in the textile-apparel industry. Those criteria are delivery (geographic location, freight terms, trade restrictions and total order lead time), flexibility (capacity, inventory availability, information sharing, negotiability and customization), cost (supplier selling price, internal costs, ordering and invoicing), quality (continuous improvement programs, certification, customer service and the percentage of on-time shipments), reliability (feelings of trust, the national political situation, the status of the currency exchange and warranty policies).

Narasimhan et al. (2006) composed a model, for global supplier selection and order allocation and considered criteria such as direct product cost, the indirect cost of coordination, quality, delivery reliability and complexity of the supply base. Lin et al. (2007) used infrastructure, country risk, government policy, human capital and cost when evaluating global suppliers. Carter et al. (2008) analysed the low-cost countries and their capabilities. They evaluated the factors such as labour cost, work ethic, intellectual property, market attraction, delivery reliability, reliable transportation, transportation costs, government support for business, political stability, flexibility, predictable border crossing and corruption in 12 different low-cost countries using perceptual mapping. They stated that experienced purchasing managers not only consider cost but also conduct a multi-criteria evaluation of global outsourcing decisions.

Au & Wong (2008) identified four main categories of factors from the literature: cost (labour costs, material costs and transportation costs), product quality (technological capabilities, reliability and trust), time to market (geographical proximity and transportation time) and country factors, including both internal country factors (such as infrastructure and ethical issues) and external country factors (such as the political and economical situation and social, linguistic and cultural differences).

Chan et al. (2008) examined the decision variables influencing global supplier selection and identified five main criteria and 19 sub-criteria: total cost of ownership (product cost, total logistics management cost, tariffs and taxes), product quality (conformance with specifications, product reliability, quality assessment techniques and process capabilities), service performance (delivery reliability, information sharing, flexibility, responsiveness and customer responses), supplier background (technological capabilities, financial status, facilities, infrastructure and market reputation), and risk factors (geographical location, political stability and foreign policies, exchange rates and economic position, terrorism and the crime rate).

Ku et al. (2010) identified the following criteria as important to global supplier selection: cost (product price, freight costs and custom duties), quality (rejection rate, process capabilities and quality assessments), service (on-time delivery, technological support, responses to changes and ease of communication), risk (geographical location, political stability and the status of the economy). Ku et al. (2010) suggested that qualitative criteria (e.g., the characteristics of the purchased items) be considered in future research.

5. Conclusion

Manufacturers must be the forerunners to be competitive in today's global markets. That is why manufacturers must keep in touch with the dynamic requirements of the market and be receptive to reforms. An increasing proportion of raw materials and work-in-process (WIP) for manufactured products is sourced globally by multinational manufacturers in today's industries.

To become a world-class manufacturer, a firm must not only compete globally in the marketplace but also be competitive and consistent in terms of costs, technological leadership, and quality. High-quality inputs are becoming the focus of many purchasing departments.

The design of global supply chains has been a challenging optimization problem for many years. In a continuing effort to remain competitive, many firms are considering new sources for their raw materials and components, new locations for their production and distribution facilities, and new markets in which to sell their products without regard for national boundaries. The design of global supply chains has been a challenging optimization problem for many years. The current globalization of the economy is forcing firms to design and manage their supply chains efficiently on a worldwide basis.

It is well known that large enterprises no longer operate in a single market. In seeking to penetrate global markets and obtain their benefits, firms are under excess pressure to reduce the price of their products and thus their production and material costs.

According to the literature, global outsourcing is mainly analyzed in terms of cost. During the evaluation of global outsourcing, neither multi-criteria evaluations nor qualitative assessments are always made, but many decision criteria must be considered when configuring a global supply chain system. Historically, labour cost has been one of the most decisive factors in global outsourcing decision making. Recently, the rapidly changing business environment has increased pressure on decision makers to properly analyze the relevant decision criteria. Strategic decision making requires the use of tangible, intangible, strategic and operational decision criteria. There are many factors that need to be considered simultaneously, and purchasing managers need a structured method of using these criteria in their decision making.

The AHP is the widely used a multi-criteria, decision-making method used by academicians and practitioners for supplier selection and global outsourcing decision making. The AHP is based on dual comparisons of decision criteria. As the number of decision criteria increases, the complexity of the system also increases. Mathematical models are not sufficient for evaluating qualitative criteria.

One of the major factors complicating the modelling of global outsourcing decision making is "uncertainty". Exchange rate fluctuations, variable transportation times, demand uncertainty, the variability of market prices, and political instability are among the most important sources of uncertainty. An effective decision-making methodology for global outsourcing must address those uncertainties. In recent years, artificial intelligence tools such as neural networks, fuzzy logic and genetic algorithms have been increasingly used for outsourcing decision making. Those methods are more appropriate under uncertainty and can better address qualitative criteria.

6. References

Aksoy, A., & Öztürk, N. (2011). Supplier Selection And Performance Evaluation In Just-In-Time Production Environments. *Expert Systems with Applications*, Vol. 38, No. 5, pp. 6351-6359, ISSN: 0957-4174

Albino, V., & Garavelli, A.C. (1998). A Neural Network Application to Subcontractor Rating in Construction Firms. *International Journal of Project Management*, Vol. 16, No. 1, pp. 9-14, ISSN: 0263-7863

Anderson, M.C., (1997). A Primer in Measuring Outsourcing Results. *National Productivity Review*, Vol. 17, No.1, pp. 33-41, ISSN 0277-8556

Antonucci, Y.L., Lordi, F.C., & Tucker J.J. (1998). The Pros and Cons of IT Outsourcing. *Journal of Accountancy*, Vol.185, No.6, ISSN: 1945-0729

Au, K.F., & Wong, M. C. (2008). Decision Factors In Global Textile And Apparel Sourcing After Quota Elimination. *The Business Review*, Vol.9, No.2, pp. 153-157, ISSN 1540 - 1200

Barla, S.B. (2003). A Case Study Of Supplier Selection For Lean Supply By Using A Mathematical Model. *Logistics Information Management*, Vol.16, No.6, pp. 451-459, ISSN 0957-6053

Boer, L., Labro, E., & Morlacchi, P. (2001). A Review of Methods Supporting Supplier Selection. *European Journal of Purchasing & Supply Management*, Vol.7, pp. 75-89, ISSN: 1478-4092.

Boer, L., Wegen, L., & Telgen, J. (1998). Outranking Methods In Support Of Supplier Selection. *European Journal of Purchasing & Supply Management*, Vol.4, pp. 109-118, ISSN: 1478-4092

Briggs, P. (1994). Case Study: Vendor Assessment For Partners In Supply. *European Journal of Purchasing and Supply Management*, Vol. 1, No. 1, pp. 49–59, ISSN 0969-7012

Brush, T.H., Martin, C.A., & Karnani, A. (1999). The Plant Location Decision in Multinational Manufacturing Firms: An Empirical Analysis of International Business and Manufacturing Strategy Perspectives. *Production and Operations Management*, Vol.8, pp.109-132, ISSN: 1937-5956

Canel, C., & Das, S.R. (2002). Modeling Global Facility Location Decisions: Integrating Marketing And Manufacturing Decisions. *Industrial Management and Data Systems*, Vol.102, No.2, pp. 110-118, ISSN 0263-5577

Canel, C., & Khumawala, B.M. (2001). International Facilities Location: A Heuristic Procedure For The Dynamic Uncapacitated Problem. *International Journal of Production Research*, Vol.39, No.17, pp. 3975-4000, ISSN 0020-7543

Canel, C., & Khumawala, B.M. (1997). Multi-Period International Facilities Location: An Algorithm And Application. *International Journal of Production Research*, Vol.35, No.7, pp. 1891- 1910, ISSN 0020-7543

Canel, C. & Khumawala, B.M. (1996). A Mixed-Integer Programming Approach For The International Facilities Location Problem. *International Journal of Operations & Production Management*, Vol.16, No.4, pp. 49-68, ISSN 0144-3577

Carter, J.R., Maltz, A., Yan, T., & Maltz, E. (2008). How Procurement Managers View Low Cost Countries and Geographies. *International Journal of Physical Distribution & Logistics Management*, Vol.38, No.3, pp. 224-243, ISSN: 0960-0035

Chan F.T.S., & Chan H.K. (2010). An AHP Model For Selection Of Suppliers In The Fast Changing Fashion Market. *International Journal of Advanced Manufacturing Technology*, Vol. 51, pp. 1195–1207, ISSN 0268-3768

Chan, F.T.S., & Kumar, N. (2007). Global Supplier Development Considering Risk Factors Using Fuzzy Extended AHP-Based Approach. *The International Journal of Management Science*, Vol.35, pp. 417-431, ISSN 0305-0483

Chan, F.T.S., Kumar, N., Tiwari, M. K., Lau, H. C. W., & Choy, K. L. (2008). Global Supplier Selection: A Fuzzy-AHP Approach. *International Journal of Production Research*, Vol.46, No.14, pp. 3825–3857, ISSN: 0020-7543

Chang B., Chang C., & Wu C. (2011). Fuzzy DEMATEL Method For Developing Supplier Selection Criteria. *Expert Systems with Applications*, Vol. 38,pp. 1850–1858, ISSN 0957-4174

Chao, C., Scheuing, E.E., & Ruch, W.A. (1993). Purchasing performance evaluation: an investigation of different perspectives. *International Journal of Purchasing and Materials Management*, Vol.29, No.3, pp. 33–39, ISSN 1055-6001

Chen, C.T., Lin, C.T., & Huang, S.F. (2006). A Fuzzy Approach For Supplier Evaluation And Selection In Supply Chain Management. *International Journal of Production Economics*, Vol.102, pp.289–301, ISSN: 0925-5273

Choy, K.L., Lee, W.B., Lau, H.C.W., & Choy, L.C. (2005). A Knowledge Based Supplier Intelligence Retrieval System For Outsource Manufacturing. *Knowledge-Based System*, Vol.18, pp. 1-17, ISSN 0950-7051

Choy, K.L., Lee, W.B., & Lo, V. (2003). Design Of A Case Based Intelligent Supplier Relationship Management System- The Integration Of Supplier Rating System And Product Coding System. *Expert Systems with Applications*, Vol. 25, pp. 87-100, ISSN: 0957-4174

Choy, K.L., Lee, W.B., & Lo, V. (2002). On The Development Of A Case Based Supplier Management Tool For Multi-National Manufacturers. *Measuring Business Excellence*, Vol.6, No.1,pp. 15-22, ISSN : 1368-3047

Coman, A., & Ronen, B. (2000). Production Outsourcing: A Linear Programming Model For The Theory-Of-Constraints. *International Journal of Production Research*, Vol. 38, No.7, pp. 1631-1639, ISSN: 0020-7543

Dana, L.P., Hamilton, R. T., & Pauwels, B. (2007). Evaluating Offshore and Domestic Production in the Apparel Industry: The Small Firm's Perspective. *Journal of International Entrepreneurship*, Vol.5, pp. 47–63, ISSN1570-7385

Dasu, S., & Torre, J. (1997). Optimizing an International Network of Partially Owned Plants Under Conditions of Trade Liberalization. *Management Science*, Vol.43, No.3, pp. 313-333, ISSN: 0025-1909

Degraeve, Z., & Roodhooft, F. (2000). A Mathematical Programming Approach For Procurement Using Activity Based Costing. *Journal of Business Finance and Accounting*, Vol. 27, No.1&2, pp. 69-98, ISSN 1468-5957

Degraeve, Z., & Roodhooft, F. (1999). Improving The Efficiency Of The Purchasing Process Using Total Cost Of Ownership Information: The Case Of Heating Electrodes At Cockerill Sambre S.A.. *European Journal of Operational Research*, Vol.112, pp. 42-53, ISSN 0377-2217

Feng, C.M., & Wu, P.J. (2009). A Tax Savings Model For The Emerging Global Manufacturing Network. *International Journal of Production Economics*, Vol. 122, No.2, pp. 534-546, ISSN 0925-5273

Ghodsypour, S.H., & O'Brien, C. (1998). A Decision Support System For Supplier Selection Using An Integrated Analytic Hierarchy Process And Linear Programming. *International Journal of Production Economics*, Vol. 56-57, pp. 199-212, ISSN 0925-5273

Ghodsypour, S.H., & O'Brien, C. (2001). The Total Cost Of Logistics In Supplier Selection, Under Conditions Of Multiple Sourcing, Multiple Criteria And Capacity Constraints. *International Journal of Production Economics*, Vol. 73, pp. 15-27, ISSN 0925-5273

Gillett, J. (1994). Viewpoint. The Cost-Benefit of Outsourcing: Assessing The True Cost of Your Outsourcing Strategy. *European Journal of Purchasing & Supply Management*, Vol. 1, No. 1, pp. 45-7, ISSN: 1478-4092

Goh, M., Lim, J.Y.S., & Meng, F. (2007). A Stochastic Model For Risk Management In Global Supply Chain Networks. *European Journal of Operational Research*, Vol.182, pp. 164–173, ISSN 0377-2217

Gottfredson, M., Puryear, R., & Phillips, S. (2005). Strategic Sourcing From Periphery to the Core. *Harvard Business Review*, February, ISSN: 0017-8012

Hadjinicola, G.C., & Kumar, K.R. (2002). Modeling Manufacturing And Marketing Options In International Operations. *International Journal of Production Economics*, Vol.75, pp. 287–304, ISSN: 0925-5273

Harris, A., Giunipero, L.C. , Tomas, G., & Hult, M. (1998). Impact of Organizational and Contract Flexibility on Outsourcing Contracts. *Industrial Marketing Management*, Vol.27, pp. 373–384, ISSN: 0019-8501

Hong, G.H., Park, S.C., Jang, D.S., & Rho, H.M. (2005). An Effective Supplier Selection Method For Constructing A Competitive Supply-Relationship. *Expert Systems with Applications*, Vol. 28, pp. 629-639, ISSN: 0957-4174

Huchzermeier, A., & Cohen, M.A. (1996). Valuing Operational Flexibility under Exchange Rate Risk. *Operations Research, Special Issue on New Directions in Operations Management, Vol.*44, No.1, pp 100-113, ISSN 0030-364X

Jiang W., & Chan F.T.S. (2011). A New Fuzzy Dempster MCDM Method And Its Application In Supplier Selection. *Expert Systems with Applications*, Vol. 38, No.8, pp. 9854-9861, ISSN 0957-4174

Karpak, B., Kasuganti, R.R., & Kumcu, E. (1999). Multi-Objective Decision-Making In Supplier Selection: An Application Of Visual Interactive Goal Programming. *Journal of Applied Business Research*, Vol.15, No.2, pp. 57-71, ISSN 0892-7626

Kheljani, J.G., Ghodsypour, S.H., & O'Brien, C. (2007). Optimizing Whole Supply Chain Benefit Versus Buyer's Benefit Through Supplier Selection. *International Journal of Production Economics*, Vol. 121, No.2, pp. 482-493, ISSN: 0925-5273

Kotabe, M. (1998). Efficiency vs. Effectiveness Orientation of Global Sourcing Strategy: A Comparison of U.S. and Japanese Multinational Companies. *Academy of Management Executive*, Vol. 12, No. 4, pp. 107-119, ISSN: 1079-5545

Kotabe, M., Mol, M.J., & Murray, J.Y. (2008). Outsourcing, Performance, and The Role of E-Commerce: A Dynamic Perspective. *Industrial Marketing Management*, Vol.37, pp. 37–45, ISSN: 0019-8501

Kotabe, M., & Murray, J.Y. (2004). Global Sourcing Strategy and Sustainable Competitive Advantage. *Industrial Marketing Management*, Vol. 33, pp. 7– 14, ISSN: 0019-8501

Kouvelis, P., & Gutierrez, G.J. (1997). The Newsvendor Problem in a Global Market: Optimal Centralized and Decentralized Control Policies For A Two Market Stochastic Inventory System. *Management Science*, Vol.43, No.5, pp.571-585, ISSN: 0025-1909

Kremic, T., Icmeli Tukel, O., & Rom, W. O. (2006). Outsourcing Decision Support: A Survey Of Benefits, Risks, And Decision Factors. *Supply Chain Management: An International Journal*, Vol. 11, No.6, pp. 467–482, ISSN: 1359-8546

Ku, C.Y., Chang, C.T., & Ho, H.P. (2010). Global Supplier Selection Using Fuzzy Analytic Hierarchy Process And Fuzzy Goal Programming. *Quality & Quantity*, Vol. 44, No.4, pp. 623-640, ISSN: 0033-5177

Kumar, S., & Arbi, A.S. (2008). Outsourcing Strategies For Apparel Manufacture: A Case Study. *Journal of Manufacturing Technology Management*, Vol.19, No.1, pp. 73-91, ISSN 1741-038x

Kumar J., & Roy N. (2011). Analytic Hierarchy Process (AHP) For A Power Transmission Industry To Vendor Selection Decisions. *International Journal of Computer Applications*, Vol.12, No.11, pp. 26-30, ISSN 0975 - 8887

Lee, E.K., Sungdo, H., & Kim, S.K. (2001). Supplier Selection And Management System Considering Relationships In Supply Chain Management. *IEEE Transactions on Engineering Management*, Vol.48, No.3, pp. 307-318, ISSN: 0018-9391

Levy, D.L. (2005). Offshoring in the New Global Political Economy. *Journal of Management Studies*, Vol.42, No.3, pp. 685-693, ISSN 1467-6486

Liao, Z., & Rittscher, J. (2007). Integration Of Supplier Selection, Procurement Lot Sizing And Carrier Selection Under Dynamic Demand Conditions. *International Journal of Production Economics*, Vol.107, pp. 502–510, ISSN: 0925-5273

Lin, Z.K., Wang, J.J., & Qin, Y.Y. (2007). A Decision Model For Selecting An Offshore Outsourcing Location: Using A Multicriteria Method. *Proceedings of the 2007 IEEE International Conference on Service Operations and Logistics, and Informatics*, pp. 1-5, ISBN 978-1-4244-1118-4, Philadelphia, PA, USA, August 27-29, 2007

Liu, J., Ding, F.Y., & Lall, V. (2000). Using Data Envelopment Analysis To Compare Suppliers For Supplier Selection And Performance Improvement. *Supply Chain Management: An International Journal*, Vol. 5, No.3, 143-150, ISSN: 1359-8546

Liu, F.F.H., & Hai, H.L. (2005). The Voting Analytic Hierarchy Process Method For Selecting Supplier. *International Journal of Production Economics*, Vol.97, No.3, pp. 308-317, ISSN 0925-5273

Lonsdale, C. (1999). Effectively Managing Vertical Supply Relationships: A Risk Management Model for Outsourcing. *Supply Chain Management: An International Journal*, Vol. 4 No. 4, pp. 176-83, ISSN 1359-8546

Lowe, T.J., Wendell, R.E., & Hu, G. (2002). Screening Location Strategies to Reduce Exchange Rate Risk. *European Journal of Operational Research*, Vol.136, pp. 573-590, ISSN 0377-2217

Mol, M.J., Tulder, R.J.M., & Beije, P.R. (2005). Antecedents and Performance Consequences of International Outsourcing. *International Business Review*, Vol.14, pp. 599–617, ISSN 0969-5931

Monczka, R.M., & Trent, R.J. (1991). Global Sourcing: A Development Approach, *International Journal of Purchasing and Materials Management*, No. Fall, pp. 2-6.

Motwani, J., Youssef, M., Kathawala, Y., & Futch, E. (1999). Supplier Selection in Developing Countries: A Model Development. *Integrated Manufacturing Systems*, Vol.10, No.3, pp.154-161, ISSN 0957-6061

Munson, C.L., & Rosenblatt, M.J. (1997). The Impact Of Local Content Rules On Global Sourcing Decisions. *Production and Operations Management*, Vol.6, No.3, pp. 277-290, ISSN 1937-5956

Narasimhan, R., Talluri, S., & Mahapatra, S. K. (2006). Multiproduct, Multicriteria Model For Supplier Selection With Product Life-cycle Considerations. *Decision Sciences*, Vol. 37, No. 4, pp. 577-603, ISSN 1540-5915

Nooteboom,B. (1999). *Inter-Firm Alliances : Analysis and Design*. Routledge. ISBN 0-415-18154-2, USA

Perron, S., Hansen, P. , Digabel, S., & Mladenovic, N. (2010). Exact And Heuristic Solutions Of The Global Supply Chain Problem With Transfer Pricing. *European Journal of Operational Research*, Vol. 202, No.3, pp. 864-879, ISSN 0377-2217

Petroni, A., & Braglia, M. (2000). Vendor Selection Using Principal Component Analysis. *Journal of Supply Chain Management*, Vol.36, No.2, pp. 63–69

Quinn, J.B. (1999). Strategic Outsourcing: Leveraging Knowledge Capabilities. *Sloan Management Review*, Vol. 40,No. 4, pp. 9-21

Quinn, J.B., & Hilmer, F.G. (1994). Strategic Outsourcing. *Sloan Management Review*, Vol. 35, No. 4, pp. 43-55

Rajan, A.J., Ganesh, K., & Narayanan, K.V. (2010). Application of Integer Linear Programming Model for Vendor Selection in a Two Stage Supply Chain. *International Conference on Industrial Engineering and Operations Management*, Dhaka, Bangladesh, January 9-10, 2010.

Ramanathan, R. (2007). Supplier Selection Problem: Integrating DEA With The Approaches Of Total Cost Of Ownership And AHP. *Supply Chain Management: An International Journal*, Vol. 12, No.4, pp. 258-261, ISSN: 1359-8546

Ray, A., Sarkar, B., & Sanyal, S. (2008). A Holistic Approach For Production Outsourcing. *Strategic Outsourcing: An International Journal*, Vol.1, No.2, pp. 142-153, ISSN 1753-8297

Ren, J., Yusuf, Y. Y., & Burns, N. D. (2009). A Decision-Support Framework For Agile Enterprise Partnering. *The International Journal of Advanced Manufacturing Technologies*, Vol. 41, pp.180–192, ISSN 0268-3768

Saen, R.F. (2007). Suppliers Selection In The Presence Of Both Cardinal And Ordinal Data. *European Journal of Operational Research*, Vol. 183,pp. 741–747, ISSN: 0377-2217

Sarkis, J., & Talluri, S. (2002). A Model For Strategic Supplier Selection. *Journal of Supply Chain Management*. Vol.38, No.1, pp.18-28

Songhori, M.J., Tavana, M., Azadeh, A., & Khakbaz, M.H. (2011). A Supplier Selection And Order Allocation Model With Multiple Transportation Alternatives. *International Journal of Advanced Manufacturing Technology*, Vol.52, No.1-4, pp. 365–376, ISSN 0268-3768

Tam, F.Y., Moon, K.L., Ng, S.F., & Hui, C.L. (2007). Production Sourcing Strategies and Buyer-Supplier Relationships: A Study of the Differences Between Small and Large Enterprises in the Hong Kong Clothing Industry. *Journal of Fashion Marketing and Management*, Vol. 11, No.2, pp. 297-306, ISSN: 1361-2026

Tan, B. (2001). On Capacity Options In Lean Retailing. *Harvard University, Center for Textile and Apparel Research, Research Paper Series*. February.

Teng, S.G., & Jaramillo, H. (2005). A Model For Evaluation And Selection Of Suppliers In Global Textile And Apparel Supply Chains. *International Journal of Physical Distribution & Logistics Management*, Vol.35, No.7/8, pp. 503-523, ISSN 0960-0035.

Timmermans, K. (2005). The Secrets of Successful Low-Cost Country Sourcing. *Accenture,* Vol.2, pp. 62-72

Ting, S.C., & Cho, D.I. (2008). An Integrated Approach For Supplier Selection And Purchasing Decisions. *Supply Chain Management: An International Journal,* Vol. 13, No.2, pp. 116-127, ISSN : 1359-8546

Vidal, C.J., & Goetschalckx, M. (2001). A Global Supply Chain Model With Transfer Pricing And Transportation Cost Allocation. *European Journal of Operational Research,* Vol.129, pp. 134-158, ISSN 0377-2217

Vidal, C.J., & Goetschalckx, M. (2000). Modeling the Effect of Uncertainties on Global Logistics Systems. *Journal of Business Logistics.* Vol.21, No.1, pp. 95-120, ISSN 0197-6729

Vokurka, R., Choobineh, J., & Vadi, L. (1996). A Prototype Expert System For The Evaluation And Selection Of Potential Suppliers. *International Journal of Operations & Production Management,* Vol. 16, No.12, pp. 106-127, ISSN: 0144-3577

Wang, G., Huang, S.H., & Dismukes, J.P. (2004). Product-Driven Supply Chain Selection Using Integrated Multi-Criteria Decision-Making Methodology. *International Journal of Production Economics,* Vol. 91, pp. 1-15, ISSN: 0925-5273

Wang, H.S., & Che, Z.H. (2007). An Integrated Model For Supplier Selection Decisions In Configuration Changes. *Expert Systems with Applications,* Vol.32, pp. 1132–1140, ISSN 0957-4174

Wang, H.S. (2008). Configuration Change Assessment: Genetic Optimization Approach With Fuzzy Multiple Criteria For Part Supplier Selection Decisions. *Expert Systems with Applications,* Vol. 34, pp. 1541-1555, ISSN: 0957-4174

Wang, J.J, Hu, R.B., & Diao, X.J. (2008-a). Developing an Outsourcing Decision Model Based on ELECTREI Method. *4th International Conference on Wireless communications, Networking and Mobile Computing,* pp. 1-4, ISBN 978-1-4244-2107-7, Dalian, October 12-14, 2008.

Wang, J.J., Gu, R., & Diao, X.J. (2008-b). Using A Hybrid Multi-Criteria Decision Aid Method For Outsourcing Vendor Selection. *4th International Conference on Wireless communications, Networking and Mobile Computing,* pp. 1-4, ISBN 978-1-4244-2107-7, Dalian, October 12-14, 2008.

Weber, C.A., Current, J.R., & Benton, W.C. (1991). Vendor Selection Criteria And Methods. *European Journal of Operational Research,* Vol. 50, pp. 2-18,I SSN 0377-2217

Wei, S., Zhang, J., & Li, Z. (1997). A Supplier Selecting System Using A Neural Network, *IEEE International Conference on Intelligent Processing Systems,* pp. 468-471, ISBN: 0-7803-4253-4, Beijing China, October 28-31, 1997.

Wu, D. (2009). Supplier selection: A Hybrid Model Using DEA, Decision Tree And Neural Network. *Expert Systems with Applications,* Vol.36, pp. 9105-9112, ISSN: 0957-4174

Xia, W., & Wu, Z. (2007). Supplier selection with multiple criteria in volume discount Environments. *The International Journal of Management Science,* Vol. 35, pp. 494-504, ISSN: 0305-0483

Yang, S., Yang, J., & Abdel-Malek, L. (2007). Sourcing With Random Yields And Stochastic Demand: A Newsvendor Approach. *Computers & Operations Research,* Vol.34, pp.3682 – 3690, ISSN: 0305-0548

Zeng, A.Z. (2000). A Synthetic Study of Sourcing Strategies. *Industrial Management & Data Systems*, Vol.100, No.5, pp.219-226, ISSN: 0263-5577

4

Reconfigurable Tooling by Using a Reconfigurable Material

Jorge Cortés, Ignacio Varela-Jiménez and Miguel Bueno-Vives
Tecnológico de Monterrey, Campus Monterrey
México

1. Introduction

Changes in manufacturing environment are characterized by aggressive competition on a global scale and rapid changes in process technology; these require creation of production systems easily upgradable by themselves and into which new technologies and new functions can be readily integrated (Mehrabi et al, 2000). In USA; industry, government and other institutions have identified materials and manufacturing trends for 2020 (Vision 2020 Chemical Industry of The Future, 2003 & National Research Council, 1998). The materials Technology Vision committee, in the publication of "Technology Vision 2020-The U.S. Chemical Industry" has identified a number of broad goals, which are enclosed in five main areas:

- New materials
- Materials characterization
- Materials modeling and prediction
- Additives
- Recycling

An important point in this vision is the development of smart materials, which have properties of self-repair, actuate and transduce. Polymers, metals, ceramics and fluids with these special characteristics belong to this class of materials and are already used in a great diversity of applications (Vision 2020 Chemical Industry of The Future, 2003).

In the other hand, the Visionary Manufacturing Challenges (National Research Council, 1998), published by the US National Academy of Sciences, presented six Grand Challenges:

- Integration of Human and Technical Resources
- Concurrent Manufacturing
- Innovative Processes
- Conversion of Information to Knowledge
- Environmental Compatibility
- Reconfigurable Enterprise

To reach these challenges, innovative processes to design and to manufacture new materials and components along with adaptable, integrated equipment, processes, and systems that can be readily reconfigured for a wide range of customer requirements or products, features, and services are needed (National Research Council, 1998).

The field of smart materials and structures is emerging rapidly with technological innovations appearing in engineering materials, sensors, actuators and image processing (Kallio et al, 2003). One of the smart materials is Nickel – Titanium alloy (NiTi) that possess an interesting property by which the metal 'remembers' its original size or shape and reverts to it at a characteristic transformation temperature (Srinivasan & McFarland, 2001).

Next manufacturing system generation requires of reconfigurable systems which go beyond the objective of mass, lean and flexible manufacturing systems. Because of the manufacturing trends towards a customer focused production. A reconfigurable manufacturing system is designed in order to rapid adjustment of production capacity and functionality, in response to new circumstances, by rearrangement or change of its components (Mehrabi et al, 2000). As can be seen in Fig. 1, there are many aspects of reconfiguration, such as, configuration of the product system, reconfiguration of the factory communication software, configuration of new machine controllers, building blocks and configuration of modular machines, modular processes, and modular tooling. So that, the development and implementation of key interrelated technologies to achieve the goals of reconfigurable manufacturing systems are needed.

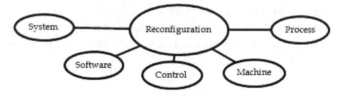

Fig. 1. Aspects of reconfiguration (Mehrabi et al, 2000).

Of relevant importance are the control, monitoring and sensing of reconfigurable manufacturing systems. By noting that the system configuration changes, the parameters of the production machines and some other physical parameters will change accordingly. The controller and process monitoring systems should have the ability to reconfigure and adapt themselves to these new conditions (Mehrabi et al, 2000).

1.1 Research justification

The use of NiTi requires proper characterization according to the environment surrounding the material when it is applied in some device; due to this requirement, a constitutive model is needed in order to relate the microstructure and thermo-mechanical behavior of the material.

In the manufacture industry, a variable shape die has always been an attractive idea to reduce design time and costs, since it allows as many designs to be rapidly manufactured at nearly free cost, using one tool for several shapes (Li et al, 2008).

The use of NiTi as an actuator in manufacturing systems is an opportunity area, allowing that several products can be formed by the same tool; this way, NiTi will help to evolve the traditional manufacturing industry.

1.2 Research aim

To develop a reconfigurable manufacture system for sheet metal/plastic forming controlled by NiTi actuators and to formulate a constitutive model of its thermo- mechanical behavior.

2. Constitutive model

NiTi is a smart material with properties such as shape memory effect (SME) and superelasticity (Chang & Wu, 2007). SME involves the recovery of residual inelastic deformation by raising the temperature of the material above a transition temperature, whereas in superelasticity, large amounts of deformation (up to 10%) can be recovered by removing the applied loads (Azadi et al, 2007).

The microscopic mechanisms involved in SME are strongly correlated to the transformation between the austenite parent phase at high temperatures and the martensite at low temperatures (Lahoz & Puértolas, 2004). It is a reversible, displacive, diffusionless, solid–solid phase transformation from a highly ordered austenite to a less ordered martensite structure (McNaney et al, 2003). Austenite has a body centered cubic lattice while martensite is monoclinic. When NiTi with martensitic structure is heated, it begins to change into the austenitic phase. This phenomenon starts at a temperature denoted by A_s, and is complete at a temperature denoted by A_f. When austenitic NiTi is cooled, it begins to return to its martensitic structure at a temperature denoted by M_s, and the process is complete at a temperature denoted by M_f (Nemat-Nasser et al, 2006). Because austenite is usually higher in strength than martensite, a large amount of useful work accompanies the shape change. Austenite exhibits higher stiffness than martensite (De Castro et al, 2007).

When NiTi is stressed at a temperature close to A_f, it can display superelastic behavior. This stems from the stress-induced martensite formation, since stress can produce the martensitic phase at a temperature higher than M_s, where macroscopic deformation is accommodated by the formation of martensite. When the applied stress is released, the martensitic phase transforms back into the austenitic phase and the specimen returns back to its original shape (Nemat-Nasser et al, 2006). The stress-induced austenite-martensite transformation is effected by the formation of martensitic structures which correspond to system energy minimizers (McNaney et al, 2003) as result of the need of the crystal lattice structure to accommodate to the minimum energy state for a given temperature (Ryhänen, 1999).

Shaw explains in more detail martensite behavior, affirming that due to its low degree of symmetry, the martensite exists either as a randomly twinned structure (low temperature, low stress state) or a stress-induced detwinned structure that can accommodate relatively large, reversible strains. Fig. 2 shows the thermomechanical response of a wire specimen. The specimen is first subjected to a load/unload cycle at low temperature, leaving an apparent permanent strain. The material starts in a twinned martensite (TM) state and becomes detwinned (DM) upon loading. The specimen is then subjected to a temperature increase while holding the load. The SME is seen as the strain is recovered and the material transforms to austenite (A). The temperature is then held at high value and the specimen is again subjected to a load/unload cycle. In this case the material shows superelasticity and transforms from austenite to detwinned martensite during loading and then back to austenite during unloading (Shaw, 2002).

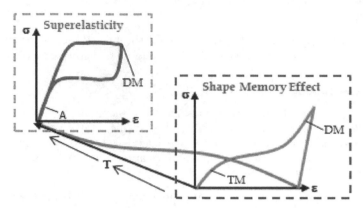

Fig. 2. Thermomechanical cycle of NiTi (Shaw, 2002)

It is considered that composition (Nemat-Nasser et al, 2006) and heat treatments have effect on the temperature at which material exhibits SME, called transformation temperatures (TTR) which are the prerequisite for the material to exhibit the SME and are one of the key parameters for SME based actuation, they also define the proper application for a certain NiTi composition alloy (Malukhin & Ehmann, 2006).

Establishment of a constitutive equation for phase transformation in NiTi requires considering the Stress-Strain-Temperature behavior and the phase transformations shown in Fig. 2, from which is observed that volume fraction of each microstructure depends of the strain and temperature conditions; it also has influence on the mechanical behavior of the material.

The possible phase transformations that can occur on NiTi are shown in Fig. 3, strain induces detwinned martensite while temperature increase induces austenite. As shown in Fig. 2, at low temperature and low stress, transformation of twinned martensite into detwinned martensite is started, and continues its plastic strain and then stress is released. When temperature is increased detwinned martensite transforms into austenite. If high temperature is kept and strain is applied superelasticity occurs and transformation of austenite into detwinned martensite is started, it finishes when stress is released and then microstructure transforms back into austenite.

According to Fig. 3, phase transformations on NiTi are:

1. Twinned martensite to Detwinned martensite (Strain induced)
2. Detwinned martensite to Austenite (Temperature increase induced)
3. Austenite to Detwinned Martensite (Strain induced)
4. Twinned martensite to Austenite (Temperature increase induced)
5. Austenite to Twinned Martensite (Temperature decrease induced)

Up to day, several studies have been made about NiTi, but there is a lack in the development of numerical analysis of the phenomenology of the material since its application requires a proper characterization; a constitutive model is needed in order to relate microstructure and thermo-mechanical behavior of NiTi. A similar model was developed by Cortes (Cortes et al, 1992) for determining the flow stress of aggregates with phase transformation induced by strain, stress or temperature and demonstrated its use for stainless steels. This model has also been applied on shape memory polymers (Varela et al, 2010) and it has been extended for modeling the displacement on electroactive polymers (Guzman et al, 2009) through a phase transformation approach, induced by some external stimulus. The model is based on an energy criterion which defines the energy consumed to deform the phases in the system as being equivalent to energy consumed to deform the aggregate and it is able to predict the flow stress behavior of the material. In order to apply Cortes' constitutive model on SMAs, experiments have to be carried out; the austenite, twinned martensite and detwinned martensite are considered as the aggregates and the microstructural transition between them becomes the basis of the constitutive model; this way, the constitutive expression will result in terms of the mechanical properties of each phase and its volume fraction.

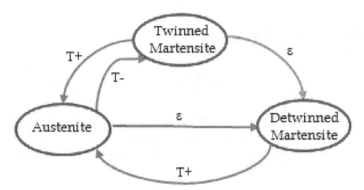

Fig. 3. Phase Transformations on NiTi and their induction stimulus.

2.1 Constitutive model of flow stress

In the case of the present aggregate composed of austenite and martensite, based on Cortes model (Cortes et al, 1992), V_f of each structure or aggregate are defined as:

$$V_{fa} = \frac{V_a}{V_t} \quad V_{ftm} = \frac{V_{tm}}{V_t} \quad V_{fdm} = \frac{V_{dm}}{V_t} \tag{1}$$

where subscripts a, tm and dm indicate austenite, twinned martensite and detwinned martensite, respectively . Cortes constitutive model of flow stress for multi phases aggregate (Cortes et al, 1992) applied on NiTi is

$$\sigma_{NiTi} = V_{fa} \cdot \sigma_a + V_{ftm} \cdot \sigma_{tm} + V_{fdm} \cdot \sigma_{dm} \tag{2}$$

where σ_{NiTi} is stress of NiTi and σ_a, σ_{tm} and σ_{dm} are stress of each structure.

2.2 Kinetics of strain/temperature induced twinned martensite-detwinned martensite-austenite phase transformation

Based on the thermomechanical behavior of Fig. 2 and the phase transformations shown in Fig. 3, volume fraction of the microstructures varies by:

$$V_{fa} + V_{ftm} + V_{fdm} = 1 \tag{3}$$

$$V_{fa} = \left(1 - V_{fa-dm}\right) \cdot V_{fa_0} + V_{ftm-a} \cdot V_{ftm_0} + V_{fdm-a} \cdot V_{fdm_0} \tag{4}$$

$$V_{fdm} = \left(1 - V_{fdm-a}\right) \cdot V_{fdm_0} + V_{ftm-dm} \cdot V_{ftm_0} + V_{fa-dm} \cdot V_{fa_0} \tag{5}$$

where subscripts 0 indicate the initial volume of each volume fraction.

For strain induced detwinned martensite phase transformation:

$$V_{fdm} = \left[1 + \left(\frac{\varepsilon}{\varepsilon_c}\right)^{-B}\right]^{-1} \tag{6}$$

For temperature induced austenite phase transformation:

$$V_{fa} = \left[1 + \left(\frac{T}{T_c}\right)^{-B}\right]^{-1} \tag{7}$$

where B is a fitting constant; while ε_c and T_c represent the values of strain and temperature, respectively at which 50% of the phase transformation is occurred. Experimental work is required for determining these values for each phase transformation. Substituting (6) and (7) in (4) and (5):

$$V_{f_a} = \left\{1 - \left[1 + \left(\frac{\varepsilon}{\varepsilon_{C3}}\right)^{-B_3}\right]^{-1}\right\} \cdot V_{f_{a0}} + \left[1 + \left(\frac{T}{T_{C4}}\right)^{-B_4}\right]^{-1} \cdot V_{f_{tm0}} + \left[1 + \left(\frac{T}{T_{C2}}\right)^{-B_2}\right]^{-1} \cdot V_{fdm_0} \tag{8}$$

$$V_{f_{dm}} = \left\{1 - \left[1 + \left(\frac{T}{T_{C2}}\right)^{-B_2}\right]^{-1}\right\} \cdot V_{f_{tm0}} + \left[1 + \left(\frac{\varepsilon}{\varepsilon_{C1}}\right)^{-B_1}\right]^{-1} \cdot V_{f_{tm0}} + \left[1 + \left(\frac{\varepsilon}{\varepsilon_{C3}}\right)^{-B_3}\right]^{-1} \cdot V_{f_{a0}} \tag{9}$$

where subscripts 1, 2, 3 and 4 represent the B, T_c or ε_c value corresponding to that phase transformation.

2.3 Stress of microstructures

Since NiTi contains a heterogeneous microstructure under given conditions, an incremental change test to determine hardening parameters in a given structure has to be carried out. Based in Cortes work (Cortes et al, 1992) flow stress of austenite, twinned martensite and detwinned martensite is determined under isothermal conditions, by pre-straining NiTi wires at a temperature at which only one microstructure exists, and then the specimens were individually deformed at a predefined temperature. The yielding point in reloading is registered as the flow stress at that temperature and those strain conditions. From this experiment equations for estimating σ_a, σ_{tm} and σ_{dm} are determined. These should be of the form:

$$\sigma_a = K_a \cdot \varepsilon^{N_a} \tag{10}$$

$$\sigma_{tm} = K_{tm} \cdot \varepsilon^{N_{tm}} \tag{11}$$

$$\sigma_{dm} = K_{dm} \cdot \varepsilon^{N_{dm}} \tag{12}$$

where K and N represent material constants which are determined experimentally.

By substituting equations (3) and (8)-(12) into equation (1) stress on NiTi can be described relating thermomechanical behavior with microstructure.

3. Reconfigurable die

Conventional type of mold fabrication involves time and money investment to achieve design of a die; the concept about forming a die of variable shape has always been attractive as a means of rapid iterations and almost cost free (Li, 2008). Thus, multi forming methods have been developed in order to achieve reconfigurability of the process.

3.1 Multi point forming (MPF)

This method has been used to replace solid dies with three-dimensional surfaces. The main key of the MPF is the two matrices of punches allowing that create a three-dimensional surface which forms according to the shape of the design; this way the surface can be approximated to a continuous die, as shown in Fig. 4 (Zhong-Yi, 2002).

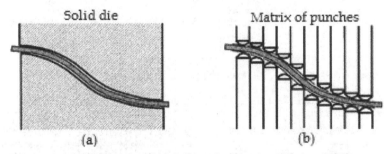

Fig. 4. (a) Conventional die forming; (b) Multi-point forming (Zhong, 2002).

MPF is based on controlling the elements and punches, hence, a matrix on punches can be shaped as required (Li, 2002). Fig. 5. illustrates the parameters related to the reconfigurable process: design and manufacture of the pin heads; since its shape, size and length play an important role in the arrangement of the closed matrix (Walczyk, 1998).

Design of a tool based on multi-point technique has many considerations since it involves several variables and many issues and problems use to occur such as dimpling, buckling and non linear deformation of the material; due to this issues four main designs have been researched with different punches types (Li, 2002).

- Multi-point full die
- Multi-point half die
- Multi-point full press
- Multi-point half press

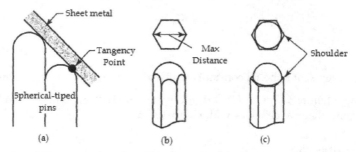

Fig. 5. Spherical and Hexagonal pin head designs (Li, 2002).

The arrangement of each design is described and shown in Table 1 and Fig. 6, respectively.

Type of Punch	Adjustment	Required Force	Drawing/ Mark
Fixed	Before Forming	Small	
Passive	While Forming	None	
Active	Free Movement	Large	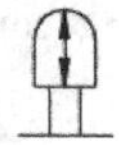

Table 1. Different types of MPF.

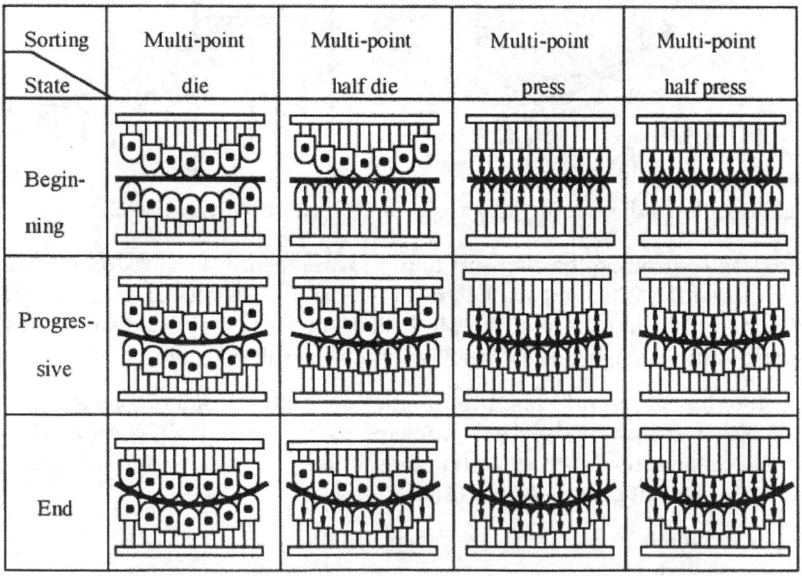

Sorting State	Multi-point die	Multi-point half die	Multi-point press	Multi-point half press
Begin-ning				
Progres-sive				
End				

Fig. 6. Different types of Multi-point forming and the interaction with the process (Walczyk, 1998).

3.1.1 Multi-point sandwich forming (MPSF)

MPSF is an accessible method to manufacture components in small batches. Fig. 7 represents the MPSF method which uses an interpolator material to assure the surface quality of the metal sheet, this last also depends of the tool elements and the position between the pins (Zhang, 2006).

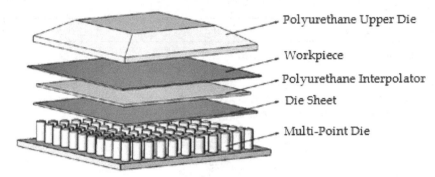

Fig. 7. Schematic Components for MPSF (Zhang, 2006).

3.1.2 Digitized die forming (DDF)

With DDF, forming procedures and integration between parameters such as deformation path, sectional forming; punches and control loop are being developed in order to avoid forming defects. The process is shown in Fig. 8 (Li, 2007).

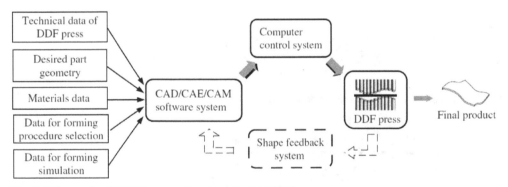

Fig. 8. Schematic of DDF integration system (Li, 2007).

3.2 Active multi point forming parameters

3.2.1 Pin head

A pin head must be strong enough to support the mechanical and thermal loads of the material to manufacture avoiding the problems in the final piece such as bending, buckling and dimpling (Walczyk, 1998). In addition to normal or vertical load, the pin will be subjected to lateral load depending the height and of the adjacent pins on the shape; this variable is a key factor on the pin head design (Schwarz, 2002), as shown in Fig. 9.

Uniformity in the pin heads and elements make the design easy to fabricate and assembly in the arrangement of the matrix, a comparison of different geometries evaluating the cross-sectional area shape is shown in Table 2 (Schwarz, 2002).

Pin Cross-Section Shape	Equilateral Triangle	Square	Hexagonal	Circle
Number of Sides	3	4	6	6
No. Isolated straight load paths	0	2	3	0

Table 2. Comparison of Cross-Sectional Geometry.

Also the structure and the size of the pins affect the quality of the piece and it is recommended the use of square shape elements, as shown in Fig. 10, in dense packed arrangements of matrices (Schwarz, 2002).

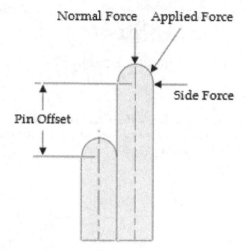

Fig. 9. Pin head forces interaction and offset.

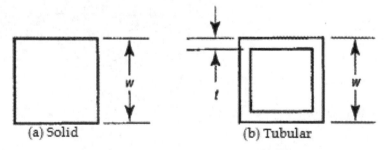

Fig. 10. Square design a) Solid and b) Tubular (Schwarz, 2002).

The use use of tubular elements is considered if the weight of the die has to be reduced, however it is not always is the best approach considering the scale and size of pins and the forces of the process (Schwarz, 2002).

3.2.2 Actuators

Each design of a shape has different means of independently moving pins in a large matrix arrangement (Walczyk, 2000). Automation of pin setting was not realized until in 1969 Nakajima positioned a matrix of pins controlled by a vibration mechanism mounted to a three-axis servomechanism (Nakajima, 1986). It is shown in Fig. 11.

Researches from different groups such as, the Massachusetts Institute of Technology (MIT) developed a Sequential Set-up Concept, the Rensselaer Polytechnic Institute build up a Hydraulic Actuation Concept and the Northrop Grumman Group Corporation created a Shaft-driven Lead screw Concept (Walczyk, 2000)

Sequential Set-up (SSU) by MIT

Each hollow pin has a threaded nut passed from its base. The pin moves up and down as the lead screw rotates, and the next pin prevent the rotation movement; the design eliminates the matrix external clamping force to position the pin heads.

Hydraulically Actuation (HA) by Renselaer Polytechnic

Individual elements are essentially hydraulic actuators, controlled by an in-line servo valve. The hydraulic pressure makes the element rise from the initial position maintaining the height until the pressure is released.

Shaft Driven Lead screw (SDL) by Northrop Group

This method depending on the need of the process needs a single or dual electric motor (one each on opposing sides of the die) mounted externally to drive worms mounted on cross shafts; the worm gear is connected to each pin´s lead screw.

Table 3 shows a comparison between these designs.

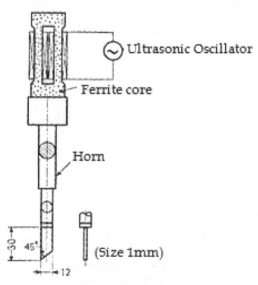

Fig. 11. Nakajima servomechanism (Nakajima, 1986).

Characteristic	SSU	HA	SDL
Matrix of Pins	42x64 (28.6mm pin)	48x72 (25.4mm pin)	42x64 (28.6mm pin)
Number of Actuators	19 (16 drive motors, X, Y and Z axes)	1 (Hydraulic pump)	42 (drive motor per row)
Number of position control devices	0	3456 (servo valve per pin)	2688 (clutch per pin)
Number of sensors	19	1	42 (encoders per motor)
Potential mayor of positioning error	Backlash in lead screw	Insufficient platen stiffness	Rotational compliance
Setting mode	Serial	Parallel	Parallel
Potential control mode error	Lead screw is continuously engaged	Moving pins are in contact with platen	Clutch does not slip
Concept Design			

Table 3. Comparison of Actuation schemes for Dies (Walczyk, 2000).

As shown in table 3; the use of common actuators have problems to form a continuous even surface, due to the size of the actuators the element diameter has to be at least 25.4mm, also the actuators and the relationship in a potential control mode error by mechanical characteristics such as fatigue and cycles limit the use of the die in order to make different shapes (Walczyk, 2000).

3.2.3 Matrix arrangement

Design and manufacturability of the pins impact the position and design of the matrix due to the number of sides on the cross-sectional shape geometry. Fig. 12 shows the key factor of densely packed pin heads in a matrix; allowing management of the load and maintaining a smoother surface when subjected to loads. The contact elements have to be maximized while the gaps between elements need to be minimized (Schwarz, 2002).

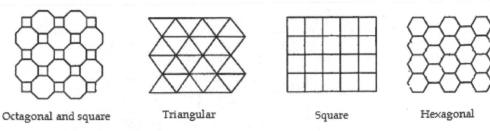

Octagonal and square   Triangular   Square   Hexagonal

Fig. 12. Various Cross-Sectional shapes for die pins (Schwarz, 2002).

Shown in Fig. 12 is a discrete digitalized arrangement to a continuous surface with multi point forming die technology. It is important to have as many pins as possible since a poor transition surface may result in delicate wrinkling defect and possible cracking of the work piece (Peng, 2006).

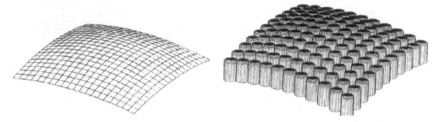

Fig. 13. Discrete approximation to a continuous surface square position (Rao, 2002).

3.2.4 Control and software

The desired part to be manufactured generating a controlling data in height of the piece is sent to a control system to perform the DDF (Li, 2007). Fig. 15. shows the different technologies merging to make the software design and control available to a reconfigurable design. An open loop control system and electronic on a single pin element, are used to evaluate a simpler circuit and timed software to excite the actuator (Walczyk & Hardt, 1998), as shown in Fig. 14.

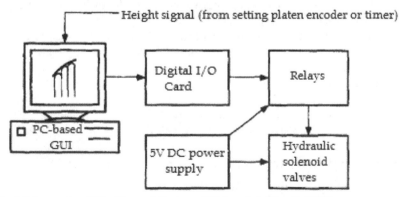

Fig. 14. Test Schematic of Die Control System (Walczyk & Hardt, 1998).

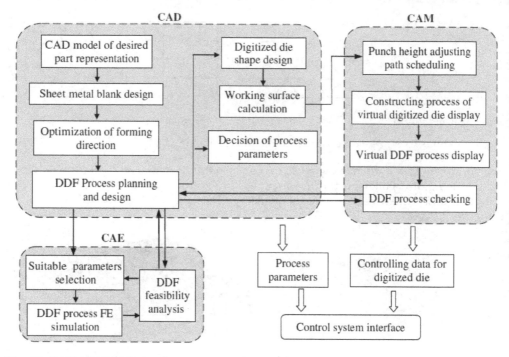

Fig. 15. CAD/CAE/CAM software control for DFF (Li, 2007).

3.3 Shape memory alloy actuators

The uses of SMA as actuators nowadays are not only medical devices; there are also novelty actuators such as (Humbeeck, 1999):

- Fashion and gadgets: single products were created from cell phone antennas, eye glasses frames and in the clothing industries frames for brassieres and wedding dresses pericoats (Duering, 1990).
- Couples: Heat-recoverable couplings of an F-14 hydraulic turbine were the first large scale produce actuator (Duering, 1990).
- Micro-actuators: The central Research Institute of Electric Power Industry in Japan built a piston-driver from a 2mm diameter wire based on 26 bars, having a life cycle over 500 000 cycles minimum.
- Adaptive materials: A vibrator frequency control of a polymer beam has been used to increase the natural frequency of the composite beam.
- Other applications: Wear cavitations defects where hydraulic machinery are used like water turbines, ship propellers and sluice channels (Jardine et al, 1994).

3.4 Development of reconfigurable die based on NiTi actuator

An active multi point forming tool, based on NiTi wires as main actuators is developed. The devices is known as 'reconfigurable die'.

3.4.1 Hypothesis

The main issue on the development of multi point surfaces is that a high density matrix of pins is required, as smaller are the pins smoother is the surface that can be formed. This issue can be solved by using a small actuator that allows formation of a dense pins matrix. SME of NiTi can be applied to achieve the movement of a mechanism that controlls the movement of each pin

3.4.2 Methodology

Development of a reconfigurable die follows the concept of DDF. Hence, it is required to design a mechanism, a controller and a graphical user interface (GUI), the full system is shown in Fig. 16.

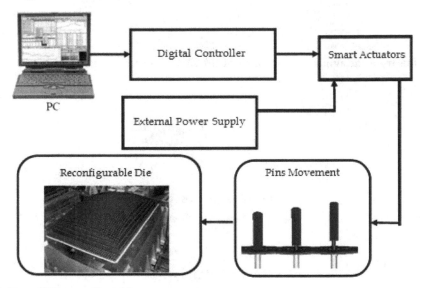

Fig. 16. Overall Process Variables.

3.4.3 Mechanism design

A design of a reconfigurable die proposed has been reviewed in order to identify its components and characterize them, sucha as the length of the shafts, springs parameters and SMA properties (length, diameter, electric current).

3.4.4 Functional prototype description

Fig 18 shows the mechanism that controls the vertical movement of each square pin. Each pin has a SMA wire subjected to a spring that deforms it, thus, NiTi has a martensitic structure. When a electric pulse activates the electric current the wire will be heated reaching the austenitic structure returning the wire to the non deformed position pushing the springs and the shaft, that causes that the pin rotates and move up; when the pulse is inactive the springs will deforms the SMA again and the cylce is restarted.

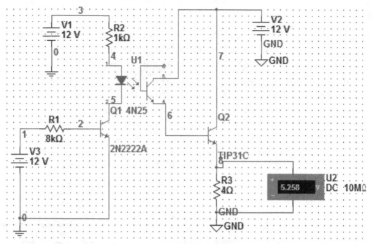

Fig. 17. Electronic design. Power and control circuits, R3 represents NiTi wire.

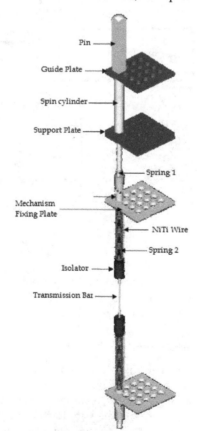

Fig. 18. General Design of the reconfigurable die.

3.4.5 Circuitry

The design consist in a basic control and power electronic circuitry which supplies the current needed to the SMA. Circuit is illustrated in Fig. 17.

3.4.6 Software

It was programmed with *Labview* by *National Instrument* and consists of three set of parts. The first one is the image codification from a solid figure to a virtual 3D figure. The second part is the pin head elements array and virtual configuration in the software; finally the last part is the digital output of electric pulses needed per pin head to change its height.

Image codification is performed by pictures taken from the object of interest, in order to make a virtual solid image in a 40 by 40 matrix. The number of views of the object depends of the geometry, a maximum of three pictures can be uploaded (sideway, front, and top views) in order to arrange the pins matrix.

The file from the generated matrix is then placed in the next file, resulting of a specific height for each element; this is then saved in a file as an array in order to visualize the pin design and making a blank. Each element has a resolution of a 28.5 pixel per pin. Fig 19 shows the Human Machine Interface (HMI) software and Fig. 20 shows the matrix of the pins height loaded on the software.

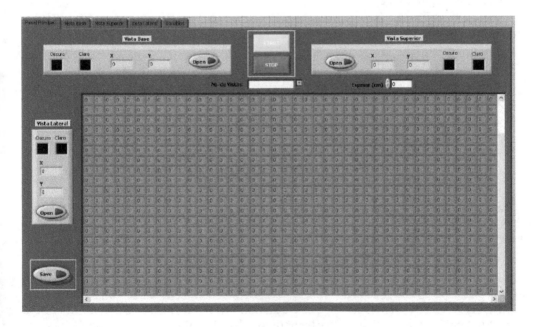

Fig. 19. Software HMI.

Fig. 20. Matrix pin height.

3.4.7 Control

Digital pulses: A digital pulse, as shown in Fig. 21, consists in a square wave of direct current output, on which the duty cycle is fixed, resulting a 50% high and 50% low. In this case the "high time" the actuator will be activated with current and the "low time" the signal will be deactivated in order to cool down the actuator.

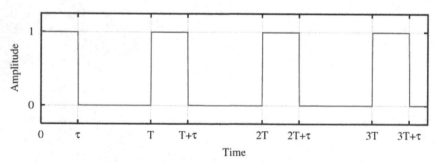

Fig. 21. Digital pulse.

The Duty Cycle (DC) represents the pulse duration divided by the pulse period, where τ is the duration that the function is actve hight (normally when the voltage is greater then zero) and T is the period of the function.

Pulse-width Modulation (PWM): is one of the most efficient ways to provide electrical power between the ranges fully on and fully off. It is a great electric tool to supply voltage/current in ht power electronics field to devices such as electric stoves, robot sensors, dimmers. The main characteristics are the variation and switching between the high and low levels in high frequencies ranges.

PWM, as shown in Fig. 22, uses a rectangular pulse wave whose pulse width is modulated resulting in the variation of the average value of the waveform. The wave has a changing duty cycle, making many pulses in a period of desired time.

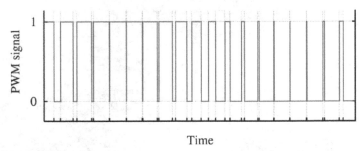

Fig. 22. PWM signal.

In order to select the most appropriate output design in the system, a Design of Experiments (DOE) it is implemented, it is shown in Table 4.

DOE		
Factor	High level	Low level
Time of pulse	4 second	2 second
Wave output	Single Pulse	PWM Pulse
Power Type	5.5 Watts	13 Watts

Table 4. DOE for evaluating the electric pulse output of each pulse.

For testing purposes, the PWM has a high frequency of 10 KHz, for the single pulse the duty cycle is a 50%. The time of pulse is the total amount of the pulse.

According to the tests, the results indicates that the factor of type of power and time of pulse have an impact and the best results on the actuator to return to the original position and having a bigger recovery force is the single pulse in a lower time with a medium range power supply.

The control design to activate the movement of the elements, there are two different methods that can be considered, as described in Table 5.

Method	Definition	Elapsed time
Serial	Elevates the height of each pin from each row is elevated one at a time.	Longer
Parallel	Elevates the height of all pins from each row is elevated.	Shorter

Table 5. Pin actuation scheme

The difference in time setting can be calculated simply by adding the quantity of total cycles from all the pins in the matrix and multiplying by the total time per one pulse in seconds.

3.4.8 Final prototype

The final alpha prototype is shown in Figs 23 and 24.

Fig. 23. Machined prototype.

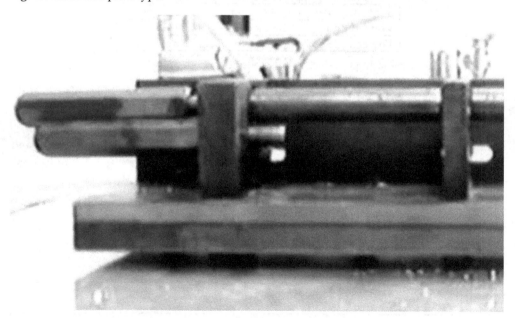

Fig. 24. Pin element changes height.

3.4.9 Operation process

The manufacture process and the use of this tool, allows a technological advantage in the design of a multi-point shape with a reconfigurable die. Its use involves CAD, CAM and CAE technologies as shown in Fig. 25.

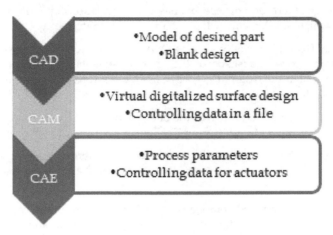

Fig. 25. Technologies used.

Image: the image depends on the views taken by the camera and the position of the object, the files are imported to the matrix arrangement software file to create a virtual object.

Virtual Matrix: when selecting the geometry of the matrix, the columns and rows are selected to visualize the object. The object then will be saved in a matrix file in the hard drive, containing the position of the elements needed from all the matrix, this allow the user to save as many designs and objects without repeating the image step every time.

Physical Matrix: Importing the matrix file, then the software will calculate the number of total cycles per element needed to reach the certain height. In this file the type of control (Parallel or Serial) is defined with the Data Adquisition Card (DAQ) digital outputs.

These proceeses are resumed in Fig. 26.

Fig. 26. Processes description.

4. Conclusions

The proposed cconstitutive model for stress on NiTi relates microstructure with thermo-mechanical behavior of NiTi. A single expression considers the 3 possible existent microstructures and their strain/temperature induced phase transformation.

The size and design of a reconfigurable tool has a strong relationship with size and design of the actuator on which the elements will be positioned in a matrix array.

The use of a reconfigurable actuator such as SMA, makes a more detail design of the pieces and decrements the size and shape of the overall elements, as seen previously in this chapter, making a highly dense pin head per matrix area resulting on a more continuous shape for the discrete shape.

The use of a step by step method such as the one proposed, makes the process an adaptable and enhanced the Vision of Manufacturing Challenges 2020 a reachable goal.

The establishment of the mechanical and electronic parameters of the proposed in order to make a functional prototype, demonstrates that the use of shape memory alloy as actuator can be possible.

The methodology recommended complies the flexibility of a reconfigurable tool according to the Manufacture Vision Challenges 2020, making a modular and adaptable process. (From the object processing image to a virtual matrix array visualizing the final surface created by the matrix arrangement of the pin elements.)

5. References

Azadi, B., Rajapakse, R., & Maijer D. Multi-dimensional constitutive modeling of SMA during unstable pseudoelastic behavior. *International journal of solids and structures*,Vol. 44, No, 20, (2007), pp. 6473-6490, ISSN 0020-7683

Chang, S., & Wu S. Internal friction of R-phase and B19' martensite in equiatomic TiNi shape memory alloy under isothermal conditions. *Journal of Alloys and Compounds*, Vol. 437, (2007), pp. 120-126, ISSN 0925-8388

Cortes, J., Tsuta, T., Mitani, Y., & Osakada, K. Flow stress and phase transformation analyses in the austenitic stainless steels under cold working. *Japan Society of Mechanical Engineers International Journal, Vol.* I35 No. 2, (1992), pp. 201-209.

De Castro J A., Melcher K., Noebe R., & Gaydosh D. Development of a numerical model for high-temperature shape memory alloys. *Smart Materials and Structures.* Vol 16. (2007). pp. 2080-2090.

Duering, TW., Melton, K., & Stockel D.Engineering Aspects of Shape Memory Alloys. *Materials and Manufacturing Processes,* (1990).

Guzman, J., Cortes, J., Fuentes, A., Kobayashi, T., & Hoshina, Y. Modeling the displacement in three-layer electroactive polymers using different counter-ions by a phase transformation approach. *Journal of Applied Polymer Science*, Vol. 112, No.6,(June 2009), pp. 3284–3293. ISSN 1097-4628

Humbeeck, J. Non-medical applications of shape memory alloys. *Materials Science and Engineering*, (1999).

Kallio, M., Lahtinen, R., & Koskinen, J. (2003). Smart materials. *Smart materials and structures*. VTT Research Program 2000-2002 Seminar. Retrieved from <http://www.vtt.fi/inf/pdf/symposiums/2003/S225.pdf>

Lahoz, R., & Puértolas, J. Training and two-way shape memory in NiTi alloys: influence on thermal parameters. *Journal of Alloys and Compounds*, Vol. 381, (2004), pp. 130–136, ISSN 0925-8388

Li, M. Multi-point forming technology for sheet metal. *Journal of Materials Processing Technology*, (2002), pp. 333-338.

Li, M. Manufacturing of sheet metal parts based on digitized-die.*Robotics and Computer-Integrated Manufacturing*, (2007), pp. 107-115.

Li, M., Cai, Z., Sui, Z., & Li, X. Principle and applications of multi-point matched-die forming for sheet metal. *Proceedings of the Institution of Mechanical Engineers, Part B: Journal of Engineering Manufacture*, Vol. 222, (May, 2008), pp. 581-589.

Malukhin K & Ehmann K. Material Characterization of NiTi Based Memory Alloys Fabricated by the Laser Direct Metal Deposition Process. *Journal of Manufacturing Science and Engineering*. Vol. 128, (2006), pp. 691-696, ISSN 1087-1357

McNaney J., Imbeni V., Jung Y., Papadopoulos P., & Ritchie R. An experimental study of the superelastic effect in a shape-memory Nitinol alloy under biaxial loading. *Mechanics of Materials*, Vol. 35 (2003), pp. 969–986, ISSN 0167-6636

Mehrabi, M., Ulsoy, G., & Koren Y. Reconfigurable Manufacturing Systems: Key to Future Manufacturing. Journal of Intelligent Manufacturing, Vol.11, No.4, (August 2000), pp. 403-419, ISSN 0956-5515

Nakajima, N. A Newly Develop Technique to Fabricate Complicated Dies and Electrodes with Wires. *Japan Society of Mechanical Engineers International Journal*, (1986).

National Research Council. *Visionary Manufacturing Challenges For 2020*, (1998). National Academies Press. Retrieved from <www.nap.edu/readingroom/books/visionary>

Nemat-Nasser S., & Guo W. Superelastic and cyclic response of NiTi SMA at various strain rates and temperatures *Mechanics of Materials*. Vol 38, (2006), pp. 463–474, ISSN 0167-6636

Peng, L. Transition surface design for blank holder in multi-point forming. *International Journal of Machine Tools & Manufacture*, (2006), pp.1336-1342.

Rao, P. A flexible surface tooling for sheet-forming processes: conceptual studies and numerical simulation. *Journal of Materials Processing Technology*, (2002).

Ryhänen J. Biocompatibility Evaluation Of Nickel-Titanium Shape Memory Metal Alloy. *University of Oulu*. (1999). Retrieved from <http://herkules.oulu.fi/isbn9514252217/html>

Schwarz, R. Design and Test of a Reconfigurable Forming Die. *Journal of Manufacturing Processes*, (2002), pp. 77-85.

Shaw, J. A thermomechanical model for a 1-D shape memory alloy wire with propagating instabilities. International Journal of Solids and Structures, Vol. 39, (2002), pp. 1275–1305, ISSN 0020-7683

Srinivasan, A., & McFarland D. *Smart Structures Analysis and Design* (2001), Cambridge University Press. ISBN 0-521-65026-7, Cambridge UK

Varela, M., Cortes, J., & Chen Y. Constitutive Model for Glassy – Active Phase Transformation on Shape Memory Polymers considering Small Deformations.

Journal of Materials Science and Engineering, Vol. 4, No. 5 (May 2010), pp 14-22, ISSN 1934-8959

Vision 2020 Chemical Industry of The Future. *Roadmap for Process Equipment Materials Technology*, (2003). Retrieved from <www.chemicalvision2020.org/techroadmaps.html>

Walczyk, D., & Hardt, D. Design and Analysis of Reconfigurable Discrete Dies for Sheet Metal Forming. *Journal of Manufacturing Systems*, Vol.17, No.6, (1998)

Walczyk, D. A Comparison of Pin Actuation Schemes for Large-Scale Discrete Dies. *Journal of Manufacturing Processes*, (2000), pp. 247-257.

Zhang, Q. Numerical Simulation of deformation in multi-point sandwich forming. *Machine Tools & Manufacturing*. (2006): pp. 699-707.

Zhong-Yi, C. Multi-point forming of three-dimensional sheet metal and the control of the forming process. *International Journal of Pressure Vessels and Piping*. (2002): pp. 289-296.

5

Towards Adaptive Manufacturing Systems - Knowledge and Knowledge Management Systems

Minna Lanz, Eeva Jarvenpaa, Fernando Garcia,
Pasi Luostarinen and Reijo Tuokko
Tampere University of Technology
Finland

1. Introduction

Today the changes in the environment, be those business related or manufacturing, are both frequent and rapid. Industry has talked about the adaptation to meet the changes over a decade. Adaptation as a word has gained quite a reputation. Adaptation is expected in design of products and processes and in the realization of processes. The adaptation in the field of manufacturing sector is commonly understood as operational flexibility and reaction speed to the changes and/or opportunities. However, in order to achieve the required level of adaptability a company must be able to learn. Learning is achieved through gaining and understanding feedback of a change: its quantity and direction. Gaining and understanding the feedback a company must be able to compare the past status to the new status of actions. Unfortunately, the knowledge of neither the past nor the present is in computer interpretable and comparable form. Thus, the achieved and/or imagined flexibility is slightly above non-existent in reality.

This chapter discusses the possibilities of a modular and more transparent knowledge[1] management concept that provides means for representing and capturing needed information as feasible as possible while understanding that it is also the software systems that need to adapt to the changes along the physical production systems. The research approach discussed here aims to introduce new ideas for the companies knowledge management and process control by facilitating the move from technology based solutions to configurable systems and processes where the digital models and modular knowledge management systems can be configured based on needs - not based on closed legacy systems. The case implementation chosen here to illustrate this approach divides the knowledge management system into three separate layers: data storing system, semantic operation logic (the knowledge representation) and services that utilize the commonly available knowledge. The modular approach in

---

[1] In the knowledge management literature, three levels of knowledge - data, information and knowledge - are commonly distinguished. Awad and Ghaziri (2004) define data as unstructured facts, which in IT terms are usually considered as just raw bits, bytes, or characters. Information is structured data and attributes which can be communicated, but which may only have meaning locked inside proprietary software. Knowledge is seen as information that has meaning for more than just one actor and it can be used to achieve results.

ICT allows also software vendors to enhance their production to be more modular and configurable thus allowing the service oriented operation model to be realized. Once the storing method is separated from the logic and services, the new concepts can emerge. It is seen also that the vendors can make new business opportunities based on modular system solutions and configuration of those instead of highly tailored solutions which cannot be re-used later on.

The chapter is structured as following: the section 2 will illustrate the challenges industrial world is facing today. Section 3 summarizes the state of the art in field of knowledge modeling. Section 4 outlines needs for modularity in systems and introduces one possible solution candidate. Section 5 introduces a case implementation. Section 6 concludes the chapter and section 7 discusses about the challenges and future trends.

2. Set of challenges for the new decade

2.1 From simple to complex operation environment

For society to sustain and prosper, it needs along with societal, structural and organizational values a steady flow of income. For most of societies manufacturing has been and still is one of the biggest source of income. However, global competition has changed the nature of European manufacturing paradigms in past decades, see Figure 1. A turbulent production environment, short product life-cycles, and frequent introduction of new products require more adaptive systems that can rapidly respond to required changes whether or not the changes are based on product design changes or changes in the production itself. However, the technological leap in the mid 20th century, provided the means to venture towards more capable systems with very highly performing components. Today the acute problem is to take full advantage of their specific capabilities. These new systems, called complex systems, are no longer reducible to simple systems like complicated ones described by Descartes, Cotsaftis (2009).

Technical developments in recent years have produced stand-alone systems where high performance is routinely reached. This solid background has allowed the extension of these systems into networks of components, which are combined from very heterogenous elements, each in charge of only a part of the holistic action of the system. As the systems are process oriented instead of knowledge oriented systems, the interaction between tasks cannot be modeled, thus the effect of single interactions and relationships cannot be represented in the full systems scale. The types of interactions are changing into a complex network of possibilities within certain limits instead of a steady and predefined process flow. This situation is relatively new and causes pressures to define the role of intended interaction. According to Chavalarias et al (2006), there is no doubt that one of the main characteristics of complex and adaptive production platforms in the future will be the ever- increasing utilization of ICT. However, while the industrial world has seen the possible advantages, the implementations fall short as a result of the required changes to the whole production paradigm, going from preplanned hierarchical systems to adaptive and self-organizing complex systems, Chavalarias et al (2006) and Cotsaftis (2009).

Chavalarias et al (2006) stated that complex systems are described as the new scientific frontier which has been advancing in the past decades with the advance of modern technology and the increasing interest towards natural systems' behavior. The main idea of the science in complex systems is to develop through a constant process of reconstructing models from constantly improving data. The characteristics of a multiple-component systems is to evolve and adapt due to internal and external dynamic interactions. The system keeps becoming a different system. Simultaneously, the connection between the system and its surroundings evolves as well. When multiple-component system is manipulated it reacts via feedback, with the manipulator and complex system inevitably becoming entangled.

Paradigm	Craft Production	Mass Production	Flexible Production	Mass Customization and Personalization	Open Complex and Adaptive Production Systems
Paradigm started	~1850	1913	~1980	2000	2020
Society needs	Customized Products	Low cost products	Variety of products	Customized products	Customized on-demand products
Market	Very samll volume per product	Steady demand	Smaller volume per product	Global manufacturing and fluctuating demand	Global manufacturing and fluctuating demand
Business Model	Pull Sell-design-make-assemble	Push design-make-assemble-sell	Push-Pull design-make-sell-assemble	Pull design-sell-make-assemble	Pull design-sell-make-assemble
Technology Enabler	Electricity	Interchangeable parts	Computers	Information technology	Information and communication technology
Process Enabler	Machine tools	Moving assembly line	Flexible Manufacturing Systems, robots	Reconfigurable Manufacturing System	Self-organizing agents

Fig. 1. Paradigm shift, adapted and modified from ManuFuture Roadmap published by European Commission (2003)

In complex systems, reconstruction is searching for a model that can be programmed as a computer simulation that reproduces the observed data 'well'. The ideal of predicting the multi-level dynamics of complex systems can only be done in terms of probability distributions, i.e. under non- deterministic formalisms. An important challenge is, contrary to classical systems studies, the great difficulty in predicting the future behavior from the initial state as by their possible interactions between system components is shielding their specific individual features. In this sense, reconstruction is the inverse problem of simulation. This naturally indicates that the complex system cannot be understood as deterministic system, since the predictions from Complex Systems Science do not say what will happen, but what can happen, Valckenaers et al (1994), Chavalarias et al (2006), Cotsaftis (2009) and Lanz (2010).

In general, complex systems have many autonomous units (holons, agents, actors, individuals) with adaptive capabilities (evolution, learning, etc), and show important emergent phenomena that cannot be derived in any simple way from knowledge of their components alone. Yet one of the greatest challenges in building a science of such systems is precisely to understand this link - how micro level properties determine or at least influence properties on the macro level. The current lack of understanding presents a huge obstacle in designing systems with specified behavior regarding interactions and adaptive features, so as to achieve a targeted behavior from the whole, Chavalarias et al (2006).

Due to the complexity of the system behavior and the lack of tangible and implementable research results on how complex systems theory can bring revenue to a company; implementations at the moment are scarce and acceptance varies. In order to meet the new requirements set by the evolving environment several new manufacturing paradigms have been introduced, which follow characteristics of natural systems. These paradigms are:

- Bionic Manufacturing System (BMS): The BMS investigates biological systems and proposes concepts for future manufacturing systems. A biological system includes autonomous and spontaneous behavior and social harmony within hierarchically ordered relationships. Cells as an example are basic units, which comprises all other parts of a biological system and can have different capabilities from each other, and are capable of multiple operations. In such structures, each layer in the hierarchy supports and is supported by the adjacent layers. The components, including the part, communicate and inform each other of the decisions, Tharumarajah et al. (1996) and Ueda et al. (1997).

- Fractal Factory (FF): The concept of a fractal factory proposes a manufacturing company composed of small components or fractal entities. These entities can be described by specific internal features of the fractals. The first feature is self-organization that implies freedom for the fractals in organizing and executing tasks. The fractal components can choose their own methods of problem solving including self-optimization that takes care of process improvements. The second feature is dynamics where the fractals can adapt to influences from the environment without a formal organization structure. The third feature is self-similarity understood as similarity of goals among the fractals to conform the objectives in each unit Tharumarajah et al. (1996) .

- Holonic Manufacturing System (HMS): The core of HMS is derived from the principles behind the term 'holon'. The term holon means something that is at the same time a whole and a part of some greater whole Koestler (1968). The model of integrated manufacturing systems consists of manufacturing system entities and related domains, the structure of individual manufacturing entities, and the structuring levels of the entities. A manufacturing system is, at the same time, part of a bigger system and a system consisting of subsystems. Each of the entities posses self-description and capability for self-organization and communication, Valckenaers et al (1994) and Salminen et al (2009).

2.2 The meaning of knowledge

It is said that the world is surrounded by knowledge. Knowledge is saved into knowledge-bases and managed by knowledge management systems is something what has been stated over and over again. However, today, no matter what the vendor flyers express with colorful pictures and highly illustrative arrows, knowledge - as computers can understand it and reason with it - is not saved. The majority of the research and design effort is never captured or re-used. The interpretation of, for example a technical drawing is entirely based on the human perception and this perception may vary. *"The meaning of knowledge is not captured and therefore not utilized as it has been intended."*

The need today is the capability for rapid adaptation to the changes in environment based on the previously acquired knowledge. However, the challenge is precisely the input knowledge or to be more accurate: the lack of it. In a large-scale company there can be up to hundreds of different design support systems, versions, and ad-hoc applications, which are used to create the information of the current product, process, and/or production systems. The majority of systems are using proprietary data structures and vaguely described semantics. This leads to challenges in information sharing since none of those are truly able to share data beyond geometrical visualizations. The design knowledge - the design intention - if even created, remains locked inside the authoring system, Ray (2004), Lanz (2010), Jarvenpaa et al. (2010), Lohse (2006) and Iria (2009).

3. The state of the art

3.1 Data modeling

As product, process and manufacturing system design have become more and more knowledge-intensive and collaborative, the need for computational frameworks to support much needed interoperability is critical. Academia and industrial world together have provided multiple different standards for product, process and resource models ranging from conceptual models to very formal representations. However, there are some serious shortcomings in the current representations:

- Firstly, none of these can represent the needs of the industry, not even industrial sector as whole.
- Secondly these standards do not form a knowledge architecture due to the missing critical parts (such as life-cycle information of products, processes and factory systems, history of past events and occurances).
- Thirdly, there does not exist a study that would outline the overlapping between these standards, Lanz et al. (2010).

Table 1 summarizes several different languages to represent data models that exist today. The list is not complete, nor it is intend to be, but it will summarize examples of standards, de facto standards and other models that are used today by industry and academia.

There have been three main approaches used to create a knowledge exchange infrastructure. They are a "point-to-point" customized solution, where dedicated interfaces are created between the design tools; a "one size fits all" solution decided by the original equipment manufacturer (OEM)'s proprietary interface for design and planning and knowledge exchange between parties; and the third solution is the a "neutral and open reference architecture" based on published standards. The first approach is expensive and time-consuming for the OEM, while the second option is very cost-efficient for the OEM, but expensive for partners who are working with several OEMs. The third option has never been fully implemented, Ray (2004), Lanz (2010) and Lohse (2006).

3.2 Knowledge capture

Second large problem area is the knowledge capturing. Currently there are very few systems that can be called knowledge capturing systems. By the definition information becomes knowledge, once other parties exist, which can understand the meaning of the information and can use it for their own purposes. In large scale organizations, data regarding activities and tasks are routinely stored in an unstructured manner, in the form of images and natural language used in e-mails, word-processed documents, spreadsheets and presentations. Over

LANGUAGE/STANDARD	USED IN PROJECTS AND STANDARDS	DESCRIPTION AND USE
EXPRESS	Standard for the Exchange of Product model data (ISO 10303 STEP), Open Assembly Model (OAM), Core Product Model (CPM), Krima et al (2009)	Defining the connections between the artifacts
CommonCADS	EUPASS Ontology, Lohse (2006)	Definition of interdependencies between classes
Web Ontology Language, Description Logics (OWL DL)/ Resource Description Framework (RDF)	Core Ontology Lanz (2010), ontoSTEP, Krima et al (2009)	Definition of interdependencies between classes and artifacts
First Order Logic (FOL)	Core Ontology, Lanz (2010)	Definition of interdependencies between classes
Common Logic Interchange Format (CLIF)	Process Specification language (PSL)	Describing what actually happens when a process specification executes and for writing constraints on processes, Bock & Gruninger (2005).
eXtensive Mark-up Language (XML)	Core Manufacturing Simulation Data (CMSD)	Used for the exchange manufacturing resource data
Automation ML	Knowledge Integration Framework ROSETTA (2010)	Representation Language of entities ROSETTA (2010)
Pabadis Promise Product and Production Process Description Language (P5DL)	OWL based language in FP6 Pabadis'Promise (2006)	P5DL used for description of products (as STEP) with their commercial and control relevant data and their necessary control applications and description of manufacturing processes with their hosting resources and necessary control functions, FP6 Pabadis'Promise (2006).

Table 1. Means for representing domain knowledge

time, large unstructured data repositories are formed, which preserve valuable information for the organization, if this information can ever be found or used.

Thus, a challenging research issue is to consider how information and knowledge is spread across numerous sources, and how it can be captured and retrieved in an efficient manner. Unfortunately, traditional information retrieval (IR) techniques not only tend to underperform on the kinds of domain-specific queries that are typically issued against these unstructured repositories, but they are also often inadequate, Iria (2009). The capturing of knowledge should start already from the creation of knowledge, where the engineer knows the meaning of the models and documents he/she is creating. This meaning should be captured in a form of computer readable format, such as a formal ontology, for further use.

3.3 Knowledge and meaning

According to DoHS (2008) increasing trend can be found from ongoing research in different domain contexts on using emerging technologies such as ontologies, semantics and semantic web (Web 2.0), to support the collaboration and interoperability. In recent years there have been a lot of activities concerning the domain and upper ontologies for manufacturing. As a result for the FP6 EUPASS project Lohse (2006) defined the connection between processes and resources for modular assembly systems. FP6 Pabadis'Promise (2006) project resulted in a manufacturing ontology (P2 ontology) and, reference architecture focusing on factory floor control.

Borgo & Leit (2007) developed the ADACOR ontology for distributed holon-based manufacturing focusing on processes and system interaction descriptions. ADACOR was later extended with an upper ontology Descriptive Ontology for Linguistic and Cognitive Engineering (DOLCE). Research done in the FP6 IP-PiSA project resulted an ontology, called Core Ontology, for connecting product, process and system domains under one reference model, Lanz (2010). The main goals these approaches generally try to achieve are: improved overall access to domain knowledge and additional information. However, none of these developed ontologies fully consider the needs above their narrow domain, Ray (2004), Lanz (2010), Jarvenpaa et al. (2010).

Ray (2004) introduced a roadmap from common models of data to self-integrating systems. The table 2 shows the 4 levels of representation. The table shows the logical steps for reaching first the creating of meaningful models (as in computational sense) to achieving finally systems that can autonomously exchange knowledge and operate based on shared knowledge.

According to the guidelines envisioned by Ray (2004), Lanz (2010) developed a common knowledge representation (KR) and semantics, called as Core Ontology, that allowed different design tools to interoperate across the design domains. The structure of the KR was formed on the basis of the requirements set by the knowledge management and integration challenges between different design tools, and the requirements set by the dynamic and open production environment. The developed model formalized the knowledge representation between product, process, and system domains utilizing fractal systems theory as a guideline. The surrounding system, be it the design environment or adaptive production system, can focus on the reasoning at different levels of abstraction, while the KR remained neutral for these

LEVELS OF REPRESENTATION	DESCRIPTION OF CHARACTERISTICS
Common Models of Data	In the lowest level the current state of the art, where the XML-based standards are utilized with relative ease within the IT sector, but not fully utilized in more conservative industry sectors.
Explicit and Formal Semantics	The second step, formal semantics, offers the generation of standardized representation that is formal enough to be parsed with computers.
Self-describing Systems	The third step is self-describing systems, where the systems can provide formal descriptions of their content and interfaces. This requires a formal semantic definition language that is rigorous enough to support logical inference.
Self-integrating Systems	The fourth level that Ray (2004) proposes is self-integrating systems. These systems are intelligent enough to be able to ask others for a description of their interfaces and, on the basis of the information thus acquired, adjust their own interfaces to be able to exchange information.

Table 2. The evolution of representational power towards formal semantics, and the systems integration capabilities that could follow (Ray, 2004)

reasoning procedures, but force the saved information to be consistent across the models. This approach differs from the traditional approaches by the fact that these tools are all utilizing already existing information as well as contributing specific information to the same model from different perspectives. The main objective of the developed KR was to achieve level two of the knowledge roadmap illustrated in the table 2. Other similar models exist, which utilize the complex nature of production systems. Most of these approaches are in the field of autonomous systems and control science.

4. Towards modular knowledge architecture for the dynamic environment

4.1 Understanding the Life-cycles of systems

All systems have their own life-cycles. In an open and complex operation environment the life-cycles play a very important role. The life-cycles that products can have are the most well-known life-cycle phases. These are such as in design, approved, in manufacturing, obsolete and such. These life-cycle phases represent the status of the design information.

The resource units also have their own specific life-cycle phases. These life-cycle phases describe the essential part of individual system units. Fore example, in the case of manufacturing resources the payload of a robot, accuracy of the tool and joints or tolerances do change over the life-cycle of the machine. It may happen that the capabilities of a system decline when it proceeds along its life-cycle. An example could be the capability for manufacturing certain surface with tight tolerances is possible when the machine is relatively new, but once the operating hours exceed a certain level the capability to reach the needed tolerances is no longer possible. Another example is the combined capability of an advanced manufacturing center and its operator. The machine may have dormant capabilities to perform advanced operations, which can be obtained once the operator has achieved needed knowledge in this particular case. Now the combined system's capabilities have increased.

In the case of modular knowledge architecture, ICT has also its own life-cycle. It is accepted from the start that the business field may change. When the change happens the ICT architecture must also adapt to the change. The change can happen also in the technological side when new technologies replace old ones. This means that some of the services may become obsolete and new services need to be added. In order to keep the architecture maintainable one solution is to offer independent service modules that operate over one information model without direct integration to the underlying databases.

4.2 Layers of operation

One of the approaches divides the knowledge management system into three separate layers: databases, semantic operation logic (the knowledge representation) and services that utilize commonly available knowledge. The modular approach in ICT allows also the software vendors to enhance their production to be more modular and configurable thus allowing the service oriented operation model to be realized. Once the storing method is extracted from the logic and services, the new concepts can emerge. It is also seen that vendors can make new business strategies based on new modular system solutions and configuration of those instead of highly tailored solutions, which cannot be re-used later on.

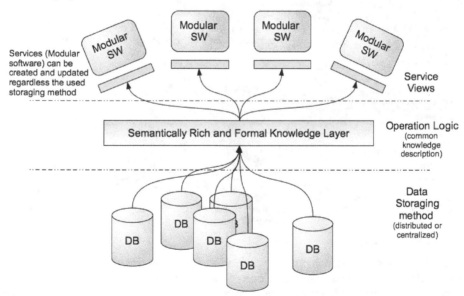

Fig. 2. Modular ICT

The ultimate goals in this particular research effort were to provide an information architecture, which allows different utilization of domain knowledge, while keeping the core information consistent and valid throughout the life-cycles of that particular set of information. The primary requirements that were defined together with industry are:

1. The model needs to represent the function of products and systems;
2. The model needs to connect different domains under one representation;
3. It must contain the history of changes applied to different instances;
4. The model must serve as an input source for automated information retrieval and reasoning in the traditional and in holon-based operation environment;
5. The model must be independent of the database implementation and services;
6. The model must allow as well as facilitate the generation of different services; and
7. The model must be extendable without disrupting the validity and consistency of the core domains.

5. Implementation of a modular ICT system

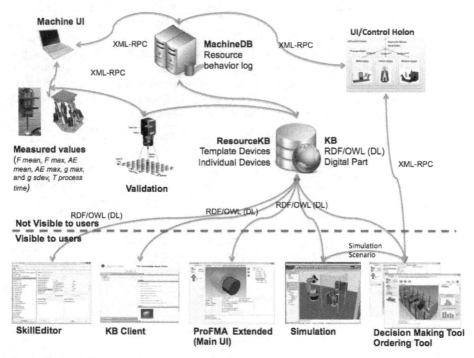

Fig. 3. implementation

The developed system, used here as an example, was based on the common knowledge representation and modular services would look as illustrated in figure 3. The clients contributing to the knowledge base are both commercial and university built existing systems and beta versions. Each of these tools requires specific domain related information and by processing the information they provide a set of services. However, the core of the system, the Knowledge Base (KB), needs to be extended to allow the capture and storing of semantically richer knowledge Lanz (2010) and Jarvenpaa et al. (2011).

The utilized knowledge representation (KR) can capture the meaning of classes via relationships that are defined between the classes. This technology allows semantic richness to be embedded into the model. Several service providers can use the meaning of stored information for their own specialized purposes. The model is divided into three separate layers as illustrated in figure 2. By dividing the data reserves, operation logic and services into separate layers connected with interfaces the upgrading of layers becomes independent of each others. This allows services to be extended, replaced and modified throughout their life-cycles.

In this case study the whole system architecture, illustrated in figure 3 has several different interoperating software modules each providing one or two essential functions for the whole holonic manufacturing system. The architecture is designed in such way that each of the modules can be replaced with a new module if needed. The connection of the modules is mainly based on the shared information model, the Core Ontology, described in detail in Lanz (2010), Lanz et al. (2011) and in Jarvenpaa et al. (2011).

The tools in the environment are designed by keeping the modularization principles in mind. Each of the tools are contributing their specific information to the common information model. The tools provide one or two main functionalities to the software environment. The modular design of the software allows changes to be applied to the tools with minimum disturbances. For example the holon user interface (UI), which controls the actual production can be replaced with a commercial tool that provides queueing functionality for the system.

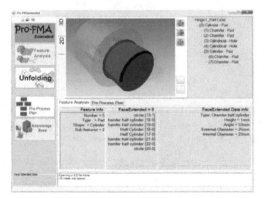

Fig. 4. Pro-FMA tool

The tools are:

Content creation: Pro-FMA illustrated in figure 4 is used to define the product requirements from the product model given in virtual reality modeling language (VRML) or eXtensive 3D (X3D) format. Product requirements are those product characteristics or features that require a set of processes for product to be assembled or manufactured. Features can be geometrical or non-geometrical by nature. These processes are executed by devices and combination of devices possessing adequate functional capabilities, Garcia et al. (2011).

Context creation: The Capability Editor, illustrated in figure 5, allows user to add devices to the ontology and assign them capabilities and capability parameters and enables creating associations between the capabilities. In other words it creates rules about which simple capabilities are needed to form combined capabilities, Jarvenpaa et al. (2011).

Fig. 5. Editor for Capabilities

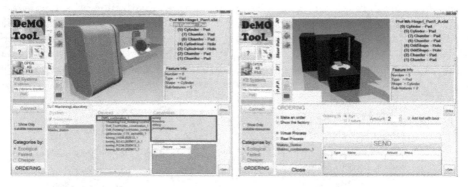

Fig. 6. Decision Making and Ordering Tool

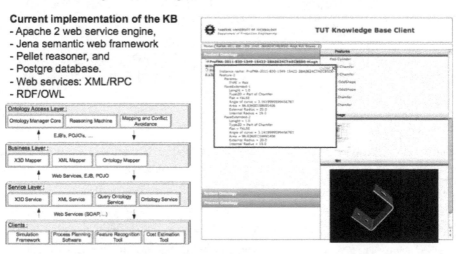

Fig. 7. Knowledge Base and Knowledge Base Web Client

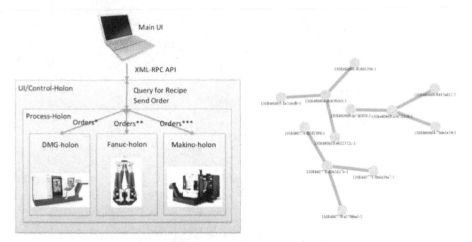

Fig. 8. Holonic control based on Kademlia, the right side of the figure shows the messages sent between holons

Verification: A simulation tool is used for creating the manufacturing or assembly scenarios. Since the environment is holonic by nature, it is accepted that the simulation only expresses possible solutions. The operation principle inside the simulation also follows holonic guidelines. This means that part or product is routed to the first available and capable cell.

Ordering: The Decision Making and Ordering Tool (DeMO tool), in figure 6, is used for setting up orders in this environment. The tool supports the viewing of the simulation as its minor function. The main function of the DeMO tool is to verify the connection to the factory floor and forward the orders to the holonic UI, Garcia et al. (2011).

Common Knowledge Representation: The KB and ResourceKB, shown in figure 7, store the information created by Pro-FMA, Capability Editor and DeMO tool. This system serves also as reference architecture, since it can handle closed models as references. The knowledge representation used in this case is based on OWL DL. The simulation model can be attached to product definition if needed. Similarly closed sub-programs and Computer-Aided-x (CAx) models can be associated with the part/product/resource description, Lanz (2010).

Content Verification: A web-based KB client, shown in figure 7 is used for human friendly information browsing. This tool serves as product data management (PDM) system's web-based user interface (UI). The client allows only limited set of changes to be applied tot he ontology. These changes are for example a new name for a product, part or other instance. For more details, please see Lanz (2010).

Operations Management: The process flow and distribution of tasks to each manufacturing or assembly cell is done with the Control Holon, see figure 8. The control holon observes the status of the system and available capabilities of system units (manufacturing resources in this case). The manufacturing resources can enter to and leave from the network without disturbing the whole system. This holonic control system distributes the tasks to suitable and available cells or stations based on the capability requirements defined by Pro-FMA earlier.

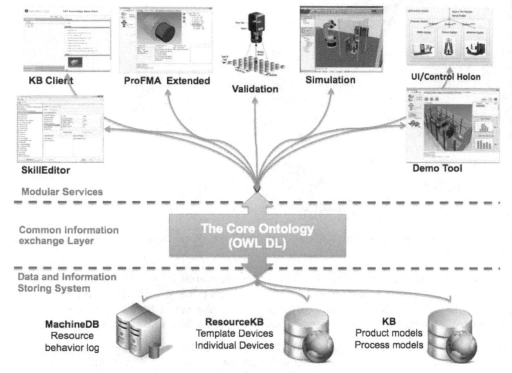

Fig. 9. The implementation is formed based on the modular ICT concept

The tools are divided into the layers described in previous chapter, see figure 2. The implemented environment, in figure 9, allows the addition of new services which can contribute and /or utilize already existing information, thus proving the modular ICT concept feasible for an adaptive, open and complex manufacturing environment. These tools constitute the necessary core for a modular system. There are additional services that could be added to this environment. These are traditional operation management module for production orchestration and machine vision based validation module. Both services are seen as extra for the core system.

6. Conclusion

Manufacturing after all is the backbone of each and every society, and in order for a society to be sustainable in long run the manufacturing has to be sustainable as well. From another point of view, manufacturing systems are shifting from being to becoming. This means that as the intelligence and cooperativeness advances the system will become a society where the rules, possibilities and constrains of a society as we know it will also apply. In order to achieve goals in the manufacturing society this research effort will contribute tremendous assets for securing the paradigm shift while keeping the manufacturing industry sustainable, flexible and adaptive. Without acceptance, further concept developments and implementation of the open and complex system approach the industry will not meet the challenges of the evolving environment.

It is seen that one partial solution is to develop these kind of modular ICT architectures that support the evolution of systems. However, it is understood that there is a lot of developments and solutions needed, since the industry cannot adopt partial solutions. Industry will require a concept that allows several data sources to be combined under one coherent and valid representation that facilitate the design and utilization of intelligent services in open and dynamic operations environment.

This paper introduced the context and operation principles of a dynamic system, and what is needed to support this kind of system from the knowledge management perspective. The article emphasized the challenge of dynamic systems from the life-cycle perspective as well, since all of the system parts be those software or hardware have their specific life-cycle phase. The division of architecture does provide tremendous possibilities for service development in future. As a proof of concept one type of modular ICT architecture and its core tools were introduced.

These results introduced here can be utilized in other fields than manufacturing engineering as well. The field of constructed environment and urban development has already seen the potential of an open world system where the input can be delivered in formal representation and services can be created independently of each other.

7. Discussion

When discussing about the holonic concepts with different people in seminars, workshops and conferences, a common comment/question has been: "Holonic manufacturing systems were developed 20-30 years ago and they didn't work then. How could they work now?" Shortly put, the answer could be technological and methodological development of knowledge and information management. Reasoning needed in the holonic systems relies on information and knowledge. Even though the concept of holonic manufacturing has remained similar throughout the years, information technology has made huge leaps enabling the implementation of these concepts in a feasible way. The novel methods to manage and distribute knowledge, such as semantic web and web service technologies, as well as semantic knowledge management systems, have been paving the way for the successful implementation of holonic systems.

Another question, which often arises in discussions has been: "Why holons? What advantages we gain by implementing holonic architecture? The implementation seems to be a huge task." Holons are autonomous and self-describing entities having well defined interfaces and the ability to communicate and co-operate with other holons. The modularity and self-organization ability enables the holonic systems to be extendable and adaptable. New holons, be they software system modules, new manufacturing resources or human workers, can enter and leave the system without disturbing the operation of the whole system. Each holon, module, knows its own purpose and the inputs and outputs, making the operation more transparent. In a holonic system it is possible to make changes in individual modules without the need to change/re-program the whole system. Until recently the holonic paradigm has only been implemented to physical devices and immediate control architecture of those. The design, operations management and supporting ICT systems have been ignored. However, as the ICT is expected to adapt to the changes in the production environment the holonic paradigm provides operation principles for this side as well.

Manufacturing is not the only domain, where the holonic paradigm could be applied. Actually, it could be applied almost anywhere, like in a medical and logistical domains. A good example can be found from city logistics. Cities, and their design, are not centrally controlled organized systems, but they are characterized by some level of chaos and the continuous threat of the chaos to expand to other operational areas. This chaos is controlled by hierarchical control systems where the control is coming from the top. From this viewpoint chaos is always considered as a negative element. This kind of systems need always be implemented as closed systems in order to prevent chaos.

The problem here is that innovations do not happen in order and harmony. The innovation always causes temporary chaos. Hierarchical control naturally strangles innovation. Therefore, what is needed is a control system where chaos is not a matter of crisis, but a normal event the system can handle in a flexible and efficient way. This kind of control system can be called as "chaordic system" (chaos + order). "Chaordic system" is self-organizing system which can always find a new equilibrium when the situation changes. The holonic control architecture can answer to the requirements of the "chaordic system". This idea has been presented to experts in the field of city logistics with very good feedback. The experts saw significant development potential for their business in holonic architecture in ICT and following the "open system" principles. However, all of this will be just theoretical discussion unless the surrounding ICT does support the change.

8. References

Awad, E.M. & Ghaziri, H.M. (2004).*Knowledge management*, Upper Saddle River, NJ, Pearson Education Inc.

Bock, C. & Gruninger, M. (2005). PSL: A Semantic Domain for Flow Models, *Journal of Software and Systems Modeling*, 4:2

Borgo, S. & Leit, P. (2007). Foundations for a core ontology of manufacturing, *Integrated Series in Information Systems*, vol 14

Chavalarias, D.; Cardelli, L.; Kasti, J. et al., (2006).Complex Systems: Challenges and opportunities, *an orientation paper for complex systems research in fp7*, European Commission

Cotsaftis, M. (2009). A passage to complex systems, in *Complex Systems and Self-organization Modeling*, C. Bertelle, G. H. E. Duchamp, and H. Kadri-Dahmani, eds., Springer, 2009

European Commission (2003). *Working Document For The MANUFUTURE 2003 Conference*

Gruver, W. (2004). Technologies and Applications of Distributed Intelligent Systems, *IEEE MTT-Chapter Presentation*, Waterloo, Canada

Dept. of Homeland Security (DoHS), NAT, Cyber Security Division: Catalog of Control Systems Security, (2008). *Recommendations for Standards Developers*

FP6 Pabadis'Promise 2006. D3.1 Development of manufacturing ontology, project deliverable, The PABADIS'PROMISE consortium

Garcia, F.; Jarvenpaa, E.; Lanz, M. & Tuokko, R. (2011). Process Planning Based on Feature Recognition Method, *Proceedings of IEEE International Symposium on Assembly and Manufacturing (ISAM 2011)*, 25-27th of May, 2011, Tampere, Finland

Iria, J., (2009), Automating Knowledge Capture in the Aerospace Domain, *Proceedings of K-CAPÕ09*,Redondo Beach, California, USA

Jarvenpaa, E.; Lanz, M.; Mela, J. & Tuokko, R. (2010). Studying the Information Sources and Flows in a Company Ð Support for the Development of New Intelligent Systems. *Proceedings of the FAIM2010 Conference*, July 14-17, 2010 California, USA, 8 p.

Jarvenpaa, E.; Luostarinen, P.; Lanz, M.; Garcia, F. & Tuokko, R. (2011).Presenting capabilities of resources and resource combinations to support production system adaptation, *Proceedings of IEEE International Symposium on Assembly and Manufacturing (ISAM 2011)*, 25-27th of May, 2011, Tampere, Finland

Jarvenpaa, E.; Luostarinen, P.; Lanz, M.; Garcia, F. & Tuokko, R. (2011). Dynamic Operation Environment Ð Towards Intelligent Adaptive Production Systems, *Proceedings of IEEE International Symposium on Assembly and Manufacturing (ISAM 2011)*, 25-27th of May, 2011, Tampere, Finland

Koestler, A. (1968). *Ghost in the Machine*, Penguin, ISBN-13: 978-0140191929, p. 400

Krima, S.; Barbau, R.; Fiorentini, X.; Sudarsan, R. & Sriram, R.D., 2009, ontoSTEP: OWL-DL Ontology for STEP, *NIST Internal Report*, NISTIR 7561

Lanz, M.; Lanz, O.; Jarvenpaa, E. & Tuokko, R. (2010). D1.1 Standards Landscape, *KIPPcolla: internal project report*

Lanz, M. 2010. *Logical and Semantic Foundations of Knowledge Representation for Assembly and Manufacturing Processes*, PhD thesis, Tampere University of Technology

Lanz,M.; Rodriguez, R. & Tuokko,R. (2010) Neutral Interface for Assembly and Manufacturing Related Knowledge Exchange in Heterogeneous Design Environment, published as book, Svetan M. Ratchev (Ed.): *Precision Assembly Technologies and Systems*, 5th IFIP WG 5.5 International Precision Assembly Seminar, IPAS 2010, Chamonix, France, February 14-17, 2010. Proceedings. IFIP 315 Springer 2010, ISBN 978-3-642-11597-4, France

Lanz, M.; Jarvenpaa, E.; Luostarinen, P.; Tenhunen, A.; Tuokko, R. & Rodriguez, R. (2011). Formalising Connections between Products, Processes and Resources Ð Towards Semantic Knowledge Management System, *Proceedings of Swedish Production Symposium 2011*, Sweden

Lohse, N. (2006), *Towards an Ontology Framework for the Integrated Design of Modular Assembly Systems*, PhD thesis, University of Nottingham

ROSETTA Project Presentation May 2010. FP7 ROSETTA Project consortium, URL: http://www.fp7rosetta.org/public/ ROSETTA_Project_ Presentation_May_2010 _webpage.pdf

Ray, S. (2004). Tackling the semantic interoperability of modern manufacturing systems, *in Proceedings of the Second Semantic Technologies for eGov Conference*

Salminen, K.; Nylund, H. & Andersson,P. (2009). Role based Self-adaptation of a robot DiMS based on system intelligence approach, *proceedings of Flexible Automation and Intelligent Manufacturing (FAIM 2009)*, US

Tharumarajah, A.; Wells, A.J. & Nemes, L. (1996). A Comparison of the Bionic, Fractal and Holonic Manufacturing Concepts, *International Journal of Computer Integrated Manufacturing*, vol.9, no.3/1996

Ueda, K.; Vaario, J. & Ohkura, K. (1997). Modelling of Biological Manufacturing Systems for Dynamic Reconfiguration, *Annals of the CIRP*, 46/1: 343-346/1997

Valckenaers, P.; van Brussel, H.; Bongaerts, L. & Wyns, J. (1994). Holonic manufacturing execution systems, *CIRP Annals - Manufacturing Technology*, nro 54/1994

6

Consideration of Human Operators in Designing Manufacturing Systems

Namhun Kim[1] and Richard A. Wysk[2]
[1]Ulsan National Institute of Science and Technology
[2]North Carolina State University
[1]Korea
[2]USA

1. Introduction

A manufacturing system normally includes various types of automated/computer controlled system resources such as material processors (e.g., CNC machines), material handlers (e.g., robots), and material transporters (e.g., AGVs) (Joshi et al., 1995). However, in most cases, to implement fully automated systems where the human is not involved is impractical (Brann et al., 1996), because of both economic and technical reasons. Furthermore, in human-involved automated manufacturing systems, a human can act as one of the most flexible and intelligent system resources in that he or she can perform a large variety of physical tasks ranging from simple material handling to complex tasks such as inspection, assembly, or packaging (Altuntas et al., 2004). From this argument, integrating a human into the system operation is a critical aspect in the design of practical manufacturing systems.

To represent the logical flows of systems' behavior, finite state automaton (FSA), formalism for discrete event-based systems, is widely used in modeling and building a control algorithm of automated manufacturing systems. While FSA-based models can be partially well suited to represent routine human activities, the vast majority of research on control models of human-involved manufacturing systems using FSA tends to consider a human as a system component that can perform tasks without considering dynamic and perceptual conditions of system constraints on human capabilities. (Shin et al., 2006b; Shin et al., 2006c). It is desirable, therefore, to include flexible and dynamic human decision making/tasks in the control of manufacturing systems with consideration of human capabilities and the corresponding system's physical conditions in human-machine co-existing environments.

Under ideal conditions, human operators should be allowed to access all physical components capable of being manipulated in the system (Altuntas et al., 2004). In this sense, a human operator can be considered a distinctive component of the system that is capable of affecting both the logical and physical states of the system. In reality, however, the human can be restricted in affecting the system components given what is afforded (e.g., offered) (Gibson, 1979) by the task environment (e.g., a part on a conveyor may be moving too fast for a human operator to grasp it). To incorporate human capabilities into the system representation, one must consider the control opportunities offered to humans by the system environment as well as the judgment demands placed on human operators.

In this chapter, a framework to develop formalisms for human-machine co-existing manufacturing systems is introduced and illustrative examples are provided in the last section.

2. Modelling of manufacturing systems

The discrete event-based modeling formalism of FSA is introduced in section 2.1. Section 2.2 presents modeling of manufacturing system control using message-based part state graph and its extended version of including human tasks into manufacturing system operations.

2.1 Finite state automata representation of DES

The fundamental physical properties of nature are considered to be continuous in that they can be expressed using real values as time changes. As systems have become more complex, event-driven approaches have become commonplace for a variety of models. Several computer technologies employ discrete methods to control complex systems such as communication networks, air traffic control, automated manufacturing systems, and computer application programs (Cassandra and Lafortune, 1999; Zeigler, 1976). Discrete event-based system modeling is a common tool to represent physical behaviors of systems, including continuous systems that are broken into discrete models which are suitable for DES-based software applications.

One of popular formalisms used to represent the logical behavior of discrete systems is based on the theories of languages and automata. This approach is based on the notion that any discrete event system can be modeled with discrete states and an underlying event set associated with it. An automaton, formalism for discrete systems, is an atomic mathematical model for finite state automata (FSA). It consists of a finite number of states and transitions that enable the model to jump between states via predetermined rules. These jumps are incurred by transition functions. These transition functions determine which state to go to next, given the current state and a current input symbol. An FSA is an effective technique capable of representing a language according to well-defined rules, which means it is rule-based and the state of the system is tractable (Sipser, 2006).

A commonly used FSA in practice is a *Deterministic Finite Automaton* (DFA), which can be defined as a 5-tuple (Hopcroft, 2001);

$$M^{DFA} = <\Sigma, Q, q_0, \delta, F>,$$

where;

Σ is a set of input alphabets (a finite non-empty set of symbols),
Q is a finite and non-empty set of states,
q_0 is an initial state such that $q_0 \in Q$,
δ is a state transition function such that $\delta: Q \times \Sigma \rightarrow Q$, and
F is a set of final states such that $F \subseteq Q$.

For example, a representation of the 5-tuple FSA for the person-climbing-stairs system is shown in Figure 1. A transition from a lower level to an upper level occurs immediately following the action of 'climb stairs' which is an input symbol to a current state "lower level."

This model only represents the physical aspects of systems behavior without considering the resource availability, and a person's attention and capability to accomplish a specific action (e.g., climb stairs). To better model human participation, it is essential to take into account the conditions required for human actions which consist of affordances (walk-on-ability) and effectivities (capability to walk) in the systems as will be explained in Section 3.

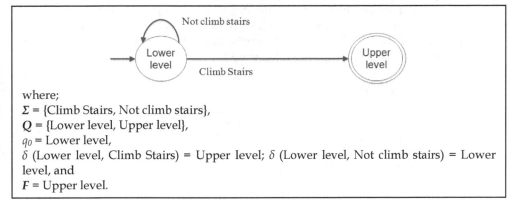

where;
Σ = {Climb Stairs, Not climb stairs},
Q = {Lower level, Upper level},
q_0 = Lower level,
δ (Lower level, Climb Stairs) = Upper level; δ (Lower level, Not climb stairs) = Lower level, and
F = Upper level.

Fig. 1. An FSA representation for the person-climbing-stairs system.

2.2 Control model of manufacturing systems

2.2.1 Message-based part state graph (MPSG)

In the 1990's, a formal model for control of discrete manufacturing systems was developed based on FSA, and called MPSG which is an acronym for Message-based Part State Graph (Smith et al., 2003). It is a modified deterministic finite automaton (DFA) similar to a Mealy machine. The MPSG model consists of sets of vertices (nodes) and edges (transitions) which correspond to the part states and the command messages, respectively. The trace of a part advancing trough an automated manufacturing system is described by its part flow diagram, which shows the sequence of part processing states in the system. As shown in Figure 2, the part state graph of a part is represented with a set of vertices and a set of edges. A vertex represents a part position in the part state graph and an edge corresponds to an operation associated with the part.

Fig. 2. An example of a part state graph for a MP class.

A MPSG describes the behavior of a controller from the parts' point of view, and each part within the domain of the controller is in a particular 'state' as described by the MPSG for that controller. The MPSG model provides no information about the system states; it determines which controller events are 'legal' with respect to that part and how to make a transition when one of these legal events occurs.

In the MPSG, all equipment level manufacturing resources are partitioned into material processors (MP; such as numerical control (NC) machines), material handler (MH; such as robots), material transporters (MT; such as automated guided vehicles (AGVs)), automated storage devices (AS), and buffer storage (BS), based on the types of their functionalities. We can create simplified physical connectivity graphs based on the MPSG controller. Figure 3 depicts a physical connectivity graph of a system, which consists of two MPs, MH, and BS, that represents physical interactions and accessibilities among the pieces of equipment. From a system's point of view, the connectivity graph is quite similar to the automaton that consists of states and transitions. However, for the individual resources, more detailed and sophisticated state transition mechanisms need to be considered, and the MPSG enables to describe the states of the entities (parts) in the system by means of the physical connectivity graph.

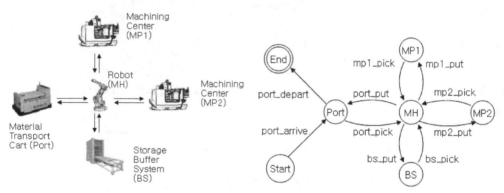

(a) Physical layout of the system.   (b) Representation of connectivity graph.

Fig. 3. Connectivity graph with two MPs, MH, BS, and port.

The MPSG M is defined formally as an eight-tuple, $M = <Q_M, q_0, F, \Sigma_M, A, P_M, \delta_M, \gamma>$, where definitions of the components are as follows:

Q_M	: Finite set of states,
$q_0 \in Q_M$	: Initial or start state,
$F \subseteq Q_M$	: Set of final or accepting states,
Σ_M	: Finite set of controller events,
A	: Finite set of controller actions,
P_M	: Physical preconditions,
$\delta_M : Q_M \times \Sigma_M \rightarrow Q_M$	: State transition function, and
$\gamma : Q_M \times \Sigma_M \rightarrow A$	: Controller action transition function.

2.2.2 Extended MPSG for human-involvement in manufacturing systems

The MPSG is a formal representation of a shop floor controller and assumes that all the resources are run in an automated way without any human involvement. To incorporate human characteristics into an automated manufacturing systems, Shin et al. investigated human-involved manufacturing systems and developed a novel formal representation by adding the tuples associated with a human element to the MPSG model (Shin et al., 2006b).

The extended MPSG model enables a human operator to cooperate with the automated pieces of equipment.

In Figure 4, solid arcs represent connections between two pieces of equipment made by automated MH equipment, whereas dotted arcs are newly created ones made by a human operator who plays as a material handler. In general, when a human operator who performs material handling tasks in a system that consists of n pieces of equipment is considered, $2 \times n$ of arcs for human transitions are created (Altuntas et al., 2004). It should be noted that the complexity of the connectivity graph increases in a linear manner.

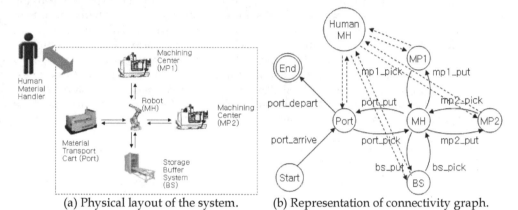

(a) Physical layout of the system.   (b) Representation of connectivity graph.

Fig. 4. Change of connectivity graphs with consideration of a human MH.

In order to express the newly created transitions by incorporating a human operator, the representation of part states is extended by incorporating information about a part location within a system and a part handling subject such that it becomes $Q = Q_M \times L \times I(p)$, where L represents a set of physical locations in the system and $I(p)$ is an interaction status with a human. In this way, an extended MPSG with a human operator, denoted by M^E, is constructed. It is defined formally as also the eight-tuple, $M^E = <Q, q^E_0, F_E, \Sigma_E, A_E, P_E, \delta_E, \gamma_E>$, where the definitions of the components are as follows:

$Q = Q_M \times L \times I(p)$: Finite set of states, where the set of state Q_M is the state of the original MPSG controller,
$q^E_0 \in Q$: Initial or start state,
$F_E \subseteq Q$: Set of final or accepting states,
$\Sigma_E = \Sigma_M \cup \Sigma_H$: Finite set of controller events, where Σ_M is a set of messages for a machine operation and Σ_H is a set of messages associated with human actions,
$A_E = A \cup \{actions\ caused\ by\ human\ activities\}$: Finite set of controller actions and human actions,
P_E : Set of Preconditions of the extended controller,
$\delta_E : Q \times \Sigma_E \rightarrow Q$: State transition function,
$\gamma_E : Q \times \Sigma_E \rightarrow A_E$: Controller action transition function,
L : Set of all physical locations in the system, and
$I(p)$: Indicator function of interaction status with a human. If a human is dealing with a part p, $I(p)=1$,. Otherwise, $I(p) =0$.

In the concept of the human-involved semi-automated system, its control depends on the complexity of a system, since a controller should recognize current status of the system and provide a proper set of commands for possible tasks based on the logical and physical preconditions. Hence, when a human material handler (human MH) performs tasks during system operation, assessment of the part flow complexity of the system needs to be conducted in developing an effective and efficient control mechanism for the system. The part flow complexity represents the possible number of tasks with a part and the possible outcomes of the tasks in terms of part states (Shin et al., 2006a).

Using this point of view, the major difference of the control schemes between the automated system and the human-involved semi-automated system is whether a human act as a passive resource of the system or a supervisory controller. The human MH can play a role as a self-regulating component which does not subordinate to the computer controller whereas other automated components perform operations in response to a given command for the controller. As such, the human MH can be considered to act as a supervisory controller, and he or she shares the system information via interfaces and sensors as shown in the Figure 5. This perspective will be further developed to expand the human's participation in complex systems.

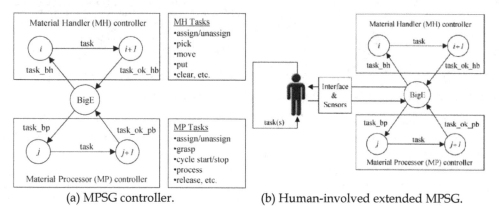

(a) MPSG controller.   (b) Human-involved extended MPSG.

Fig. 5. Control scheme of the MPSG and extended MPSG controllers (Shin et al., 2006a).

3. Modelling and control of human-machine cooperative manufacturing systems

In section 3.1, a representation of human-involvement considering prospective human action opportunities (affordance) is introduced. The modeling basis and formal control model for affordance-based human-machine cooperative manufacturing system are presented in section 3.2 and 3.3, respectively. The example of affordance-based MPSG system control with a simple and typical manufacturing cell is illustrated in section 3.4.

3.1 Human-involvement in system representation

In dynamic situations, the interactions between humans and environs play a key role in achieving an ecosystem's goal. Identifying opportunities for interactions between them is important to the modeling and operation of human-involved systems in an effective way. In this section, we introduce a formal modeling methodology that combines human actions into the system control scheme in formal mathematical FSA.

The concept of affordances implies that human-involved systems are composed of two or more related objects including at least one human and one environmental component, (an affordance complementary property consisting of the dual relationship between animals (humans) and their environs). The terms of affordance and effectivity are treated as an environmental reference and the animal's capability to take actions in the environment. In the sense of a formal representation of affordances, the environmental and animal components are combined together so that they incur a different property to be activated (Turvey, 1992).

Turvey presents a formal definition of affordances mathematically using a juxtaposition function as follows;

Let $W_{pq}=j(X_p,Z_q)$ be a function that is composed of two different objects X and Z, and further p and q be properties of X and Z, respectively. Then, p refers to an affordance of X and q is the effectivity of Z, if and only if there exists a third property r such that:
i. $W_{pq}=j(X_p,Z_q)$ possesses r,
ii. $W_{pq}=j(X_p,Z_q)$ possesses neither p nor q, and
iii. Neither X nor Z possesses r, where r is a joining or juxtaposition function.

For example, a person (Z) can walk (q), stairs (X) that can support something (p), and they together yield a climbing property (r) as shown in Figure 6. This formal definition corresponds to a mathematical formalism of an FSA in that it describes properties as discrete states, and the juxtaposition function can be mapped to the state transition function in the FSA. The existence of a formal definition of an affordance provides a foundation that the concept of an affordance can be combined with software engineering and systems theory.

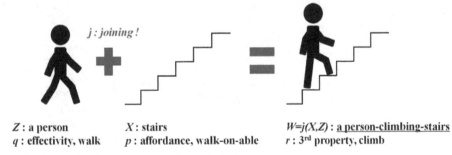

Z : a person X : stairs $W=j(X,Z)$: <u>a person-climbing-stairs</u>
q : effectivity, walk p : affordance, walk-on-able r : 3rd property, climb

Fig. 6. An example of a 'person-climbing-stairs' system.

If we regard the states of the environmental system as discrete ones and consider the transitions among the states which are triggered by possible actions of animals or other system resources, an ecosystem of an environment and humans can be represented by an FSA (Kim et al., 2010). The theory of automata corresponds to the ecological sense of affordances for at least the following two reasons: 1) an environmental system can be defined as a set of nodes and arcs which describe discrete states of the system and the transitions between states, respectively, and 2) a set of transitions between states represents a set of potential properties (affordances) of the environmental system which can be triggered by certain human activities and lead to the next states. Therefore, affordance-effectivity combinations can be considered conditions for identifying possible human actions using FSA representations.

There is a set of physically connected transitions from one state to another, which corresponds to a set of dispositional properties of affordances in the system. The set of feasible transitions is triggered if and only if the input symbol is taken as a parameter of a transition function in the environmental system. This input symbol is considered an effectivity. Next, the circumstances need to be specified in order for a human transition to occur in terms of the general representation of the FSA, $M^{DFA} = <\Sigma, Q, q_0, \delta, F>$. The conditions that allow humans to make transitions within a system can be represented by a four-tuple, $<X_p, Z_q, J, W_{pq}>$, which comes directly from Turvey's definition of affordance. By merging these two sets of tuples, an extended automaton for incorporating affordances of a system and effectivities of humans within the system can be constructed. The new representation for the formal model of affordance and effectivity in FSA is $M^{DFA'} = < \Sigma, Q, q_0, \delta, F, X_p, Z_q, J, W_{pq}>$, where;

J is a Juxtaposition function such that $J: X_p \times Z_q \rightarrow W_{pq}$,
X_p is a set of affordances in the system,
Z_q is a set of effectivities of human in the system,
W_{pq} is a set of possible human actions in the system, and
all other definitions of tuples are the same as those of M^{DFA}.

The graphical representation of the affordance-based FSA, $M^{DFA'}$, for the person-climbing-stairs system is shown in Figure 7. Transition from a lower level to an upper level occurs, *if and only if a human is able to 'walk (X_p)' and the stairs are 'walk-on-able (Z_q)' for human, which means 'Climb Stairs (system input of human action) $\in W_{pq}$.'*

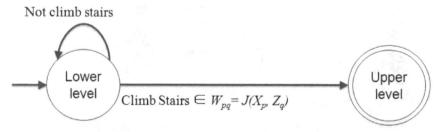

Fig. 7. Affordance-based FSA for the person-climbing-stairs system(Kim et al., 2010).

From an ecosystem's perspective, if the set of all transitions among the system states can be considered Σ. A state transition occurs only when the transition for some input alphabet is included in the set of transitions in the system, $a \in \Sigma$, where a represents an input alphabet. From the human's point of view, he or she has a set of effectivities (capabilities or available actions) regardless of the transitions included in Σ. Thus, transitions occur if and only if the set of possible human actions, W_{pq}, are executed (the results of juxtaposition between specific affordance and effectivity). Component, $h \in W_{pq} \subseteq \Sigma$, represents the possible set of actions for a human to actualize on the environmental system, causing state transitions. In this sense, the dispositional properties that come from joining the properties of affordance and effectivity are considered possible human actions on the human-environmental system. In many unstructured instances, the set of actions can be an infinite set, but for well-structured environs the set of actions can be a very small set. The relationship among affordances of a system, effectivities, and actions of human can be depicted as shown in Figure 8.

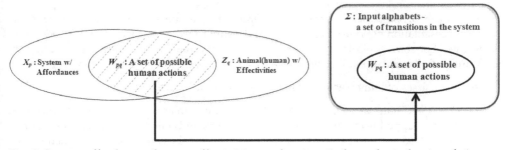

Fig. 8. System affordances, human effectivities, and actions in the ecological point of view (Kim et al., 2010).

The FSA-based modeling formalism for manufacturing systems control can take human activities into account, where source and sink state nodes are defined within the system state behaviors. However, the existing control model of human-involved manufacturing systems lacks prospective control perspectives. It only considers human operators as flexible system components acting like robots, rather than animals which have nondeterministic natures of recognitions and physical limitations. To develop the formal model of human-machine cooperative systems, the ecological sense of system affordances and human effectivities should be included in the model for the seamless control of the systems.

Special care needs to be taken for human operators since their actions are those of a nondeterministic autonomous agent that perceives, measures, and makes a judgment in the system in consideration of other resources and environmental aspects. For this reason, affordances for a human operator in the system need to be considered carefully for human-machine cooperative systems. It can then contribute to assess the human effects on the system in a more effective way.

3.2 Representation of affordances in human-involved manufacturing systems

For a formal control model of human-involved manufacturing systems, it is necessary to incorporate affordances within a system that accounts for possible human actions with regard to at least the material handling processes with consideration of physical limitations for the actions, such as size, weight, and temperature. This corresponds to distinguishing possible human actions from human capable actions (effectivities). We remark that the set of possible human actions are a part of the collection of human *potentially* capable actions considering that human may or may not take actions due to his or her cognitive recognition of actions or physical limitations imposed by an environment. From the viewpoint of the manufacturing system with a human material handler, the affordance can be described as;

Define W_{pq} as a set of *possible human actions in manufacturing system*. Let X_p be a physical state of a part or a piece of equipment in a system where p is a human accessibility (affordance) to it, and Z_q be a human material handler where q is a human action if and only if there is the third property r such that:

i. $W_{pq}=j(X_p,Z_q)$ possesses r : human material handing action (location change),
ii. $W_{pq}=j(X_p,Z_q)$ possesses neither p nor q, and
iii. Neither X nor Z possesses r, where r is a joining or juxtaposition function.

In order to incorporate system affordances into the manufacturing system controller, a formal representation of occurrences of the third properties, called dispositions, needs to be established. For some typical possible human actions for material handling (e.g., *access*, *pick*, *move*, and *put*), the corresponding circumstance can be specified as follows (Kim et al., 2010);

The specific classes of human activities to be addressed include the following:

1. For a human material handler to be able to **access** a machine (resource), the machine should be stopped before the human starts to work and other MHs (robots) should not run on it. Also, the human needs to perceive that he or she can work on the machine.
2. For a human material handler to be able to **pick** a part, at least one degree of freedom (DOF) of the part needs to be released and the human can separate the part from the position.
3. For a human material handler to be able to **move** a part, he or she should be holding the part, and the central position of the part can be changed in the global Cartesian coordinate system as a result of the *move* action.
4. For a human material handler to be able to **put** a part on a machine, the machine should be stopped and the number of parts on the machine should not exceed the capacity of the machine. Also, the machine needs to support the part (fixture without slip).

As described above, there are some obvious state transitions for a human material handler in the system. Based on the examples, we decompose human material handling tasks into four types of actions (*access*, *pick*, *move*, and *put*), and also define system affordances and human capable actions corresponding to them as follows;

<Affordances in the manufacturing system with human material handlers>

The specific classes of human activities to be addressed include the following:

1. A machine is **accessible**; the machine is stopped and waits to process a part, and no other MH is working on it. The machine volume should be within the human's access ranges.
2. A part is **pickable**; the chuck or fixture holding the part is open, and at least one DOF of the part is available. The part should weigh less than maximum lifting force, and should be less than maximum grapping width for a human material handler.
3. A part is **movable**; the part is held by a human, and the location of the part can be changed by human actions. There are no *substantial* obstacles from a starting point to an ending position of the human.
4. A part is **putable**; the machine stops working, and it can support the part upright without slip.

<Human MH's capable actions (effectivities) in manufacturing systems>

1. A human material handler can **access** a piece of equipment.
2. A human material handler can **pick** a part from a piece of equipment.

3.   A human material handler can **move** a part to a piece of equipment.
4.   A human material handler can **put** a part to a piece of equipment.

The third property in Turvey's affordance formalism is mapped on a subset of possible human actions. By doing this, the juxtaposition function can be formulated based on its definition. In the definition of the set of possible human actions, denoted by $W_{pq}=j(X_p,Z_q)$, j is the joining or juxtaposition function. If X_p and Z_q have multiple dispositions, the juxtaposition function j needs to filter p and q from the dispositions possessed by X_p and Z_q to realize the possible actions of W_{pq}. To construct a juxtaposition function to address this, X_p and Z_q are expressed as row matrices that consist of '0' and '1', which represent a certain property *exists* ('1') or *not* ('0') in the system. Thus, in a manufacturing system with human, the sets of P and Q can be expressed as following equation (1) and (2), respectively;

$$P = (Accessible,\ Pickable,\ Movable,\ Putable)\text{: Properties of the system} \tag{1}$$

$$Q = (Can\ Access,\ Can\ Pick,\ Can\ Move,\ Can\ Put)\text{: Properties of a human} \tag{2}$$

By multiplying each component in the matrices, the juxtaposition function of this problem is defined as in equation (3) to obtain the third properties and possible state transitions;

$j : X_p \times Z_q \rightarrow W_{pq}$ and $\pi : P \times Q \times C \rightarrow PA$, where P is a set of affordance status for a part state, Q is a set of action capability (effectivity) status to the human operator, PA is a set of possible human actions in the system, and C is a set of physical action conditions (preconditions for human actions).

Suppose, $P = \{p_i : i=1,2,3,4\}$, $Q = \{q_j : j=1,2,3,4\}$ where p_i and q_j are binary numbers, then

$$PA = \begin{cases} \phi & if\ p_1q_1=0. \\ \{((p_2q_2 \times 'pick'),(p_3q_3 \times 'move'),(p_4q_4 \times 'put'))\}-\{0\} & if\ p_1q_1=1\ \&\ C\ \text{is true.} \end{cases} \tag{3}$$

Note that the empty set refers to a situation that a human operator cannot access resource.

3.3 Formalism for human-machine cooperative systems: Affordance-based MPSG

As mentioned in the previous section, some human actions become available depending on the environmental affordances, and transitions made by human actions can be realized by satisfying both system affordances and corresponding human effectivities (capable actions). Affordances should have ontological assumptions related to space and time as in Gibson's ecological definition (Gibson, 1979). In the sense of system controller such as the MPSG, it is one of the key factors to build formal representation of the affordance concepts that imposing quantifiable metrics on affordances.

From the MPSG point of view, the supervisory controller, called a Big-E, has a module to generate possible transitions based on the logical validation of preconditions as shown in the Figure 9. The existing MPSG controller generates process plans based on the fully automated systems that are assumed to properly operate as planned beforehand. In this sense, as long as the system is working without *critical* failures, the human action is not necessary and a human is allowed to intervene between machine operations whenever he or she decides to do so. However, when an unanticipated incident occurs (e.g., machine down, oversized part), which is usually beyond the controller's resolution capability, human involvement is required. In this case, the Big-E controller notifies a human of the case so that a human operator can step in the process for preceding the system to the next available proper transition (Kim et al., 2010).

From the viewpoint of a human operator, he or she could make a transition in the system to move a part forward toward completion (one of the feasible ways to proceed when the system requires some human action). The set of feasible transitions are mapped into the Big-E controller, which can generate possible alternative action commands based on the logical validation modules. It is worth note that this exactly corresponds to the set of system affordances for this case. It should also be noted that not all feasible transitions are available for the human operator because the system affordances for the human operator have ontological assumptions of physical and time domains, as mentioned above.

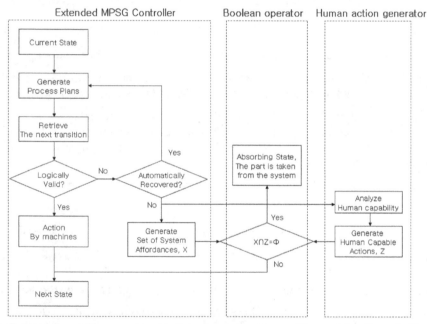

Fig. 9. Control flow of human-involved automated system with consideration of affordances (Kim et al., 2010).

In order to realize a human cooperative system in the ecological sense, generation modules for two distinctive logical sets and the Boolean operator for juxtaposing these two logical sets need to be constructed for a human operator to cooperate with the controller with consideration of affordances as shown in Figure 9.

Considering the formal representation of affordances, the extended MPSG for human-involved system control can be improved in such a way that it can consider more realistic transitions by human possible actions. In this chapter, the affordance-based MPSG, denoted by M^A, is defined as a 12-tuple, which comprises eight-tuples from the initial extended MPSG model, M^E, and four-tuple from the affordance representation. It is defined formally as $M^A = <Q, q^E_0, F_E, \Sigma_E, A_E, P_a, \delta_E, \gamma_E, X, Z, J, W >$, where the definitions of the components are as follows:

J is a juxtaposition function such that $J: X \times Z \to W$,
X is a set of affordances,
Z is a set of effectivities (human capable actions),
W is a set of possible human actions,

where;

$$J(x(p, 1), z(p, 1)) = W$$

$$= \begin{cases} \varnothing & \text{if } x_1 z_1 = 0 \\ j(X,Z) = \left\{ \begin{matrix} x_2 z_2 \times pick\_p\_from\_l_1, \\ x_3 z_3 \times move\_p\_from\_l_1\_to\_l_2, \\ x_4 z_4 \times put\_p\_on\_l_2 \end{matrix} \right\} - \{0\} & \text{if } x_1 z_1 = 1 \end{cases}$$

where;

$1 \subseteq L$, $x(p, 1) \in X$, $z(p, 1) \in Z$,
$x(p, 1) = x(p, \{l_1, l_2\}) =$ (a location set $\{l_1, l_2\}$ is accessible, part 'p' is 'pickable' at l_1, part 'p' is movable from l_1 to l_2, part 'p' is 'putable' on l_2), and
$z(p, 1) = z(p, \{l_1, l_2\}) =$ (access to a location set $\{l_1, l_2\}$, pick the part 'p' at l_1, move the part 'p' from l_1 to l_2, put the part 'p' on l_2)'
δ_E is a state transition function such that $\delta_E: Q \times \Sigma_E \to Q$.,

where;

$$\delta_E((v, l, I(p)), a, W) = (\delta_M(v, a), l, 0), \text{ if } a \in \Sigma_M \text{ (by a MH)}$$

$$\delta_E((v, l, I(p)), a, W) = \delta_H((v, l, I(p)), a, W), \text{ if } a \in W \subset \Sigma_H \text{ (by a human)}$$

$$\delta_E((v, l, I(p)), a, W) = (v, l, I(p)) \text{ if } a \notin W \subset \Sigma_H \text{ (no transition)},$$

where δ_M is a state transition function by automated MHs (robots) and δ_H is a state transition function by a human material handler, and all other definitions of tuples are the same as those of M^E.

Based on the above definition, the juxtaposition function can generate a set of possible human actions under a particular circumstance when system affordances are defined as environmental situations, time limitation, physical layout of a system, and part properties, e.g., size, volume, and speed. The human transition set of the affordance-based MPSG is a subset of that of the extended MPSG as shown in Figure 10.

Thus, the complexity of the MPSG controller of human cooperative systems can be reduced when the concept of affordances are taken into account. In the extended MPSG controller, M^E, physical preconditions, P_a, are evaluated so that some impossible transitions can be prevented. However, the physical preconditions may account for only a small part of system affordances that can be measured by pre-installed sensors, while most of possible human transitions are determined by human cognitions. It is noteworthy that system affordances for humans have much greater impact on the operations and control of the human-machine cooperative systems.

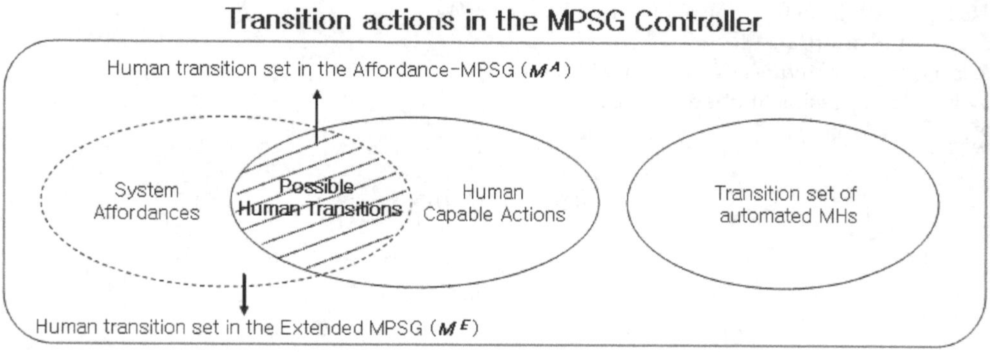

Transition actions in the MPSG Controller

Human transition set in the Affordance-MPSG (M^A)

System Affordances | Possible Human Transitions | Human Capable Actions

Transition set of automated MHs

Human transition set in the Extended MPSG (M^E)

Fig. 10. Transition action sets in the Affordance-based MPSG controller (Kim et al., 2010).

3.4 Illustrative example: Affordance-based MPSG model

This section presents an application example to illustrate the proposed manufacturing control model with affordances. As shown in Figure 11, two types of graphs are constructed to represent the system's physical configuration and the logical control logic. The first graph shows the relationship among the resources in a system and possible path for parts. Based on this connectivity graph in Figure 11(a), the affordance-based FSA representation in Figure 11(b) can be created to develop a control scheme for the system. This is then used to generate an affordance-based MPSG controller that incorporates operations of each piece of equipment and possible human actions (Kim et al., 2010).

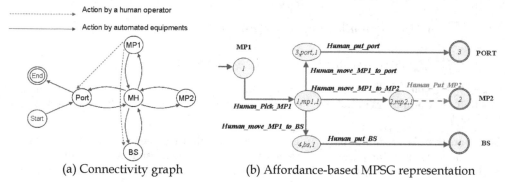

(a) Connectivity graph   (b) Affordance-based MPSG representation

Fig. 11. FSA representation; MP2 has no affordance of put-ability for human operators (Kim et al., 2010).

Specifically, Figure 11 depicts a case in which a human operator can move a part from 'MP1' to anywhere when the part is not 'putable' on 'MP2', i.e., the MP2 is located so far from the operator that he or she cannot see if the MP2 is empty. The affordance and effectivity matrices between 'node 1' and 'node 2' are expressed with the proposed model as follows,

$$x(part,\{MP1,MP2\}) = (x_1, x_2, x_3, x_4) = (1,1,0,1) \text{ and}$$

$$z(part, \{MP1, MP2\}) = (z_1, z_2, z_3, z_4) = (1,1,1,1)$$

So, the juxtaposition function can be,

$$W = \begin{cases} 1 \times pick\_part\_from\_MP1, \\ 1 \times move\_part\_from\_MP1\_to\_MP2, \\ 0 \times put\_part\_on\_MP2 \end{cases} - \{0\}$$

$$\therefore W = \{pick\_part\_from\_MP1, move\_part\_from\_MP1\_to\_MP2\}$$

If the human material handler wants to make a transition between 'MP1' and 'MP2', he or she needs to take three actions (pick, move, and put) between the nodes. However, the action, 'put', is not available in the system. It means that by taking the affordances in the system, the complexity of the graph in terms of the number of possible human actions in the FSA representation can be reduced.

The eligible affordances and effectivities of the example can be expressed as follows;

The affordance chart at time t:

$$x(part, \{MP1, MP2\}) = (1,1,0,1)$$

$$x(part, \{MP1, BS\}) = (1,1,1,1)$$

$$x(part, \{MP1, PORT\}) = (1,1,1,1)$$

The effectivity chart at this point:

$$z(part, \{MP1, MP2\}) = (1,1,1,1)$$

$$z(part, \{MP1, BS\}) = (1,1,1,1)$$

$$z(part, \{MP1, PORT\}) = (1,1,1,1)$$

From the above affordances and effectivities relationships, we obtain;

$$W = \{Pick\_part\_from\_MP1, Move\_part\_from\_MP1\_to\_MP2,$$
$$Move\_part\_from\_MP1\_to\_BS, Move\_part\_from\_MP1\_to\_PORT,$$
$$Put\_part\_on\_BS, Put\_part\_on\_PORT\}$$

If the controller is to allow a part transition between MP1 and MP2 by a human material handler, W should contain a complete set of actions which is composed of *pick*, *move*, and *put* between MP1 and MP2. However, W does not have *'put'* on MP2 actions in itself in this example. Thus, the human operator cannot make the part transit between MP1 to MP2 as a material handler.

4. Function allocation between human and machine

Dynamic task allocation control scheme for realization of human-machine cooperative systems is introduced in section 4.1. Classification of errors and their recoveries in human-machine cooperative systems using affordance-based MPSG are presented in section 4.2.

4.1 Work allocations in human-machine cooperative systems

Sheridan (2000) discusses a list to assert "what men are better at" and "what machines are better at" (MABA-MABA) as follows;

<Humans are usually conceived to be better at>:

1. Detecting small amount of visual, auditory, or chemical energy.
2. Perceiving patterns of light or sound.
3. Improvising and using flexible procedures.
4. Storing information for long periods of time, and recalling appropriate parts.
5. Reasoning inductively.
6. Exercising judgment.

<Machines are better at>:

1. Responding quickly to control signals.
2. Applying great force smoothly and precisely.
3. Storing information briefly, erasing it completely.
4. Reasoning deductively.

The gaps between 'what machines are better at and what humans are better at' are getting narrower as machines are replacing human more and more with the development of artificial intelligence technologies. However, the complete replacing humans with the automated machines are almost impossible and impractical partly because of both economic and technical reasons (Brann et al., 1996).

In this sense, the function allocations between machines and humans in the human-involved automated system are one of the vital factors to control the system in effective and flexible ways. As pointed out in the previous section 3.1, human actions are available depending on the environmental affordances, and transitions by human can be realized by satisfying both system affordances and corresponding human effectivities Thus, from the system point of view, the controller needs to differentiate the set of actions that humans are better at from actions that machine are better at with consideration of availability of human actions identified by the model.

Suppose that the material handling time (e.g., time for picking up, moving, and putting a part) and material lifting capability (e.g., part weight, volume, size, and temperature) can be critical factors to allocate work between human material handlers and robots in a manufacturing cell. If the controller is able to evaluate the availability of a resource (either human or machine) which can reduce a processing time for a material handling job at a certain point of time and space, the whole system works faster and more intelligent to increase its productivity. For example, if we consider a simple human-machine cooperative manufacturing cell shown in Figure 4 with following characteristics;

> *Time for a robot to move a part from a resource to a resource = 10 ± 0.5 sec.,*
> *Time for a human to move a part between adjacent resources = 5 ± 2 sec.,*
> *Time for a human to move a part between facing resources = 10 ± 5 sec., and*
> *Human capable part size and weight ≤ 3ft × 3ft × 3ft and 20 LB.*

The controller is able to evaluate expected average processing time for each task and allocate the task between a human operator and a robot based on information of the dynamic location of a part and the human operator, and system working status. For instance, if the human operator is waiting for a message from Big-E within three seconds walking distance from MP1, and a part, whose size and weight are 1 ft × 1 ft × 1 ft and 10 LB, needs to be moved from MP1 to BS, the expected average time of the human task to move the part is eight seconds and that of the robot is 10 seconds. Thus, the human operator is supposed to be faster than the robot to accomplish this specific task, and the controller will allocate the task to the human operator as shown in Figure 12. In this case, the external transition function in affordance-based MPSG needs to be revised as follows,

$$\delta_E : Q \times \Sigma_E \rightarrow Q$$

$\delta_E((v, l, I(\text{part})), a) = \delta_H ((v, l, I(\text{part})), a)$ if $a \in PA \subseteq \Sigma_H$ and the human is expected to perform a task faster than the automated MH, and
$\delta_E(v, l, I(\text{part})), a) = (\delta_M (v, a), l, 0)$, otherwise.

Check:
if any possible human action is available, and
whether 'human is better' or 'machine is better'

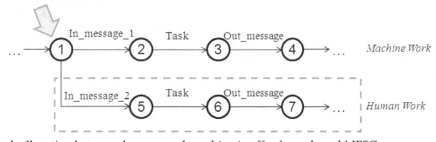

Fig. 12. Task allocation between human and machine in affordance-based MPSG.

4.2 Classification of errors

In the perspectives of systems theory and controls, a human agent is neither completely controllable nor perfectly predictable because of his or her nondeterministic and complex behaviors. For this reason, human-machine interactive system models need to harness dynamic human decision making processes into discrete system contexts. The level of modeling grains is defined with respect to the modeling purposes and modelers' perspectives on the systems. Representation of systems using finite numbers of states and transitions poses a lot of challenges to make a model complete by itself. Thus, comprehensive definition and classification of errors and error states in discrete system models can increase modeling easiness, simplicity and completeness.

For instance, the human-involved automata model of 'a semi-automated manufacturing system' illustrated in section 3.4 should contain an additional system state of the absorbing (error) state. In this modeling representation, human actions and system transitions that may not lead to the desired states, which come from the goal of the human-involved system, directly go to the absorbing state. Only valid interactions between a human and a system can be parts of a human-involved or human-machine cooperative process that change system states from a current to a next state which is placed within the process to the desired goal states.

It may not be critical to investigate errors in descriptive system representation as mentioned above. However, in the perspectives of system control models, system recoveries from errors are important to accomplish the seamless and complete modeling of human-involved systems. Thus, investigation of errors and their proper classification in systems are one of keys to develop control models for human-machine cooperative systems.

4.2.1 Error classification in extended MPSG systems

In human-involved systems, human errors are considered important factors from a control point of view because sometimes system status is significantly changed by the human errors which are not within traceable states. It is well known that there are a number of topics to be addressed in terms of human errors. Shin et al. (2006c) investigated human operational errors concerning the human material handling in extended MPSG controls. In the authors' research, only human operational errors that are directly related with physical material handling tasks are considered, and human operational errors are classified into two separate categories; location errors and orientation errors.

A location error means that a human material handler made a mistake to pick or put a part on a wrong resource location. A human may commit a location error during his or her material handling task by loading or unloading a specific part on some equipment (resources) which are not in the proper process plans for the part. An orientation error is the case of not properly placing a part on equipment. For example, a human may commit an orientation error when he or she places and fixes the part on the controllable vice. Even if the human operator places the part on the right equipment (location), he or she may make an orientation error because of placement of the part in wrong directions and fixation of the part improperly.

Location and orientation errors may hinder the system from starting a proper operation in processes, and this failure causes the system to stop and wait for a recovery action. Every part in system operations is represented by its own unique state that is specified in a part-state graph with electronic sensors that can check the physical precondition of a system operation a, $\rho_a \in P_E$. Therefore, location and orientation errors are checked by sensors installed on equipment before the system starts a process.

4.2.2 Error classification in affordance-based MPSG systems

The location and orientation errors stated in the previous section 4.2.1 are taxonomies under physical preconditions regarding coordination states of a part in systems. However, in the ecological definition of affordances, properties of affordances, effectivities, and possible human actions in systems should have ontological assumptions related with space and time (Turvey, 1992), and the failure to satisfy these assumptions can lead a system state to undesirable states or make an improper transition. The cases of failing to satisfy assumptions in space dimension fall into the category of location and orientation errors. The cases of failing to satisfy assumptions in time domains, however, were not investigated.

In the perspectives of control models, actual system status and behaviors should coincide with representation of states and events within the same time and space domains. The detection of location and orientation errors can be easily performed by using sensors installed on resources (equipment) in control systems, while the detection of failing to satisfy time constraints cannot be considered in the existing extended MPSG control systems. Specifically, the automated equipment in systems run based on the logical preconditions within systems, but a human in the system tends to take an action relying on his or her perception-based actions which are available within a specific space dimension and time duration containing the affordance-effectivity duals for those actions. For this reason, one additional error classification for a human needs to be considered; a set of transition errors with respect to time and space constraints between a human and a part. A human may commit transition errors if he or she missed to perform a desired task within a specific time range.

An example of transition errors can be expressed in affordance-based MPSGs as shown in Figure 13. The errors can be detected and checked when a specific human action is not taking within the time and space conditions described in a set of action conditions, C. The action conditions can be estimated based on the information of the relative properties between a human material handler and a part, such as size and weight of the part, lifting and moving capabilities of the human material handler, relative distance between the part and human. The size and weight of a part can be detected by sensors installed on equipment, the human capabilities are pre-programmed based on the personal information, and the location, viewing, and moving direction of the human material handler can be detected by a vision sensor installed in the shop floor system. The representation of affordance-based MPSG systems contains time-related tuples which can measure and check the time constraints for existence of possible human actions. The time advances are checked within control programs for equipment and the system allows a human material handler to perform human tasks only within a specific time range.

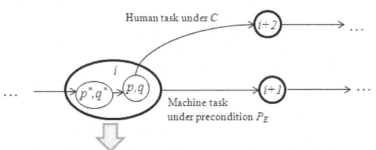

Transition error: though the action condition C is satisfied and the controller asked to perform a human task in state i, the human operator did not take any action within accessible time ranges.

Fig. 13. Examples of location and orientation errors in affordance-based MPSG systems.

4.2.3 Error recovery

The detection and classification of human errors in human-machine cooperative systems are crucial to validate the control processes of human-involved systems. The analysis of error status in systems can guarantee the prompt and proper recovery of the systems from undesirable system states.

When location and orientation errors are occurred, the system will stop and wait for recovery action by either incurring automatic recovery module or calling human material handlers. In case of a transition error, the system can simply recover it by re-allocating the human task to a machine without stopping and recalling an error recovery module as shown in Figure 14, if the desired task can be performed by either a human or a machine. If a desired human task is failed to be performed within an eligible time range, machine can take an action instead of a human. However, if the required task for a specific system transition can be done only by a human, it should be recovered by human operators. The transition error recovery process is described as shown in Figure 15.

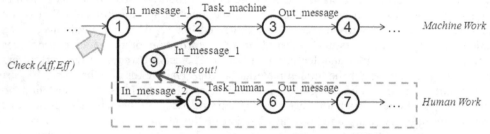

Fig. 14. Recovery of a transition error by re-allocating a human task to machine.

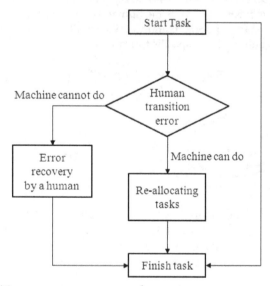

Fig. 15. Human transition error recovery procedure.

5. Summary

This chapter presents the modeling concept and formal representation of human-involved manufacturing control systems called affordance-based MPSG. With consideration of affordances in manufacturing systems, a human can participate in system operations and dynamic task allocation between a human and a machine is available.

Investigation of errors and their classification are also discussed. In regard to human transition errors, the automatic task reallocation to machine is a solution to solve the errors in easy ways. However, if the original task for a specific transition was only available for a human operator, an error recovery task by a human should be incurred to solve it.

6. References

Altuntas, B.; Wysk, R. A. & Rothrock, L. (2007). Formal approach to include a human material handler in a computer-integrated manufacturing (CIM) system. *International Journal of Production Research*, Vol. 45, No. 9, pp. 1953-1971, ISSN 0020-7543

Brann, D. B.; Thurman, D. A. & Mitchell, C. M. (1996). *Human interaction with lights-out automation: a field study*, IEEE, ISBN 0-8186-7493-8, Dayton, OH, USA

Cassandra, C. G. & Lafortune, S. (1999). *Introduction to Discrete Event Systems*, Springer, ISBN 0792386094, Boston, MA

Fitts, P. M.; Viteles, M. S.; Barr, N. L.; Brimhall, D. R.; Finch, G.; Gardner, E.; Grether, W. F.; Kellum; W. E. & Stevens, S. S. (1951). *Human engineering for an effective air navigation and traffic control system*, Ohio state University Fundation Report, Columbus, OH

Gibson, J. J. (1979). *The ecological approach to visual perception*, Psychology Press, ISBN 0898599598, Boston, Houghton Mifflin.

Hopcroft, J. E. (2001). *Introduction to automata theory, languages, and computation*, Addison-Wesley, ISBN 0201441241, Boston, MA

Joshi, S. B.; Mettala, E. G.; Smith, J. S. & Wysk, R. A. (1995) Formal models for control of flexible manufacturing cells: physical and system model. *IEEE Transactions on Robotics and Automation*, Vol. 11, No. 4, pp. 558-570, ISSN 1042-296X

Kim, N.; Shin, D. ; Wysk, R. A. & Rothrock, L. (2010) Using finite state automata (FSA) for formal modelling of affordances in human-machine cooperative manufacturing systems. *International Journal of Production Research*, Vol. 48, No. 5, pp. 1303-1320, ISSN 0020-7543

Shin, D.; Wysk, R.A. & Rothrock, L. (2006a). Formal Model of Human Material-Handling Tasks for Control of Manufacturing Systems. *IEEE Transactions on Systems, Man, and Cybernetics, Part A: Systems and Humans*, Vol. 36., No. 4, pp. 685-696, ISSN 1083-4427

Shin, D.; Wysk, R. A. & Rothrock, L. (2006b). A formal control-theoretic model of a human-automation interactive manufacturing system control, *International Journal of Production Research*, Vol. 44, No. 20, pp. 4273-4295, ISSN 0020-7543

Shin, D.; Wysk, R. A. & Rothrock, L. (2006c). An investigation of a human material handler on part flow in automated manufacturing systems, *IEEE Transactions on Systems, Man, and Cybernetics Part A: Systems and Humans*, Vol. 36, No. 1, pp. 123-135, ISSN 1083-4427

Smith, J. S.; Joshi, S. B. & Qiu, R. G. (2003). Message-based Part State Graphs (MPSG): A formal model for shop-floor control implementation, *International Journal of Production Research*, Vol. 41, No. 8, pp. 1739-1764, ISSN 0020-7543

Turvey, M. T. (1992). Affordance and Prospective Control: An Outline of the Ontology. *Ecological Psychology*, Vol. 4, No. 3, pp. 173-187, ISSN 1040-7413

Zeigler, B. P. (1976). *Theory of Modeling and Simulation*, John Wiley & Sons, ISBN 0-12-778455-1, New York, NY

7

PLC-Based Implementation of Local Modular Supervisory Control for Manufacturing Systems

André B. Leal[1], Diogo L. L. da Cruz[1,3] and Marcelo da S. Hounsell[2]
[1]*Department of Electrical Engineering, Santa Catarina State University*
[2]*Department of Computer Science, Santa Catarina State University*
[3]*Pollux Automation*
Brazil

1. Introduction

Developing and implementing control logic for automated manufacturing systems is not a trivial task. Industrial production lines should be able to produce many types of products that go through a growing number of processes given the needs of the market, and there is an ever growing flexibility demand because of it. To keep up with it a faster way to develop control logic automation for the production lines is required. And this should be done in such a way to easy development and to guarantee that the control is correct in terms of making the system to behave as it should. To this end, the use of formal modelling tools seems to help raise the abstraction level of specifying systems' behaviour at the same time that it provides ways to test the resulting model.

The Supervisory Control Theory (SCT) of Ramadge and Wonham (1987, 1989) is an appropriate formal tool for the control logic synthesis of automated systems because it ensures the achievement of an optimal control logic (minimally restrictive and nonblocking), and that meets control specifications. Regardless of its advantages for automated manufacturing systems control and troubleshooting, this theory and its extensions have not been broadly used in industrial environments so far. The main reason for this resides in some difficulties that exist in dealing with complex problems. According to Fabian and Hellgren (1998) another important reason is the difficulty in implementing a pragmatic solution obtained from SCT theoretical result, i.e., bridging practice and theory.

This chapter presents a methodology, named DECON9, that aims to reduce the gap between this promising theory and real world applications, i. e., it presents a methodology for the implementation of the SCT into Programmable Logic Controllers (PLCs). The local modular approach (Queiroz & Cury, 2000a, 2000b) is used for the supervisors' synthesis and the implementation in PLC is performed in the ladder diagram language. Local Modular approach is used because systems of greater complexity that have a big amount of machines (and then, events) usually can be modelled as many concurrently interacting and simpler subsystems.

PLC implementation of supervisory control was also discussed in (Ariñez et al., 1993; Lauzon, 1995; Leduc & Wonham, 1995; Leduc, 1996; Qiu & Joshi, 1996; Lauzon et al. 1997; Fabian & Hellgren, 1998; Dietrich et al., 2002; Hellgren et al., 2002; Liu & Darabi, 2002; Music &

Matko, 2002; Queiroz & Cury, 2002; Chandra et al., 2003; Hasdemir et al., 2004; Manesis & Akantziotis, 2005; Vieira et al., 2006, Morgenstern & Schneider, 2007; Noorbakhsh & Afzalian 2007a&b; Afzalian et al., 2008; Hasdemir et al., 2008; Noorbakhsh, 2008; Silva et al., 2008; Leal et al., 2009; Possan & Leal, 2009; Uzam et al., 2009). In most of these works the monolithic approach (Ramadge & Wonham, 1989) for the supervisors' synthesis is used, in which a single and usually large supervisor is computed to control the entire plant. According to (Queiroz & Cury, 2002), this approach is not adequate for most real problems because they involve a large number of subsystems. In order to overcome this problem, in some works the synthesis of supervisors is performed according to the local modular approach (Queiroz & Cury, 2000a), which reduces the computational complexity of the synthesis process and the size of supervisors by exploiting specifications modularity and the decentralized structure of composite plants. Thus, instead of a monolithic supervisor for the entire plant, a modular supervisor is obtained for each specification, taking into account only the affected subsystems.

In almost all these work the implementation is held on ladder diagram, which is a well-known PLC programming language in industrial environments. But most existing proposals can only tackle one event per PLC scan cycle, which represents a problem when handling large scale plants (Vieira et al. 2006). Just a few of those proposals, at the best situation, can process one event per supervisor at each PLC scan cycle, a situation that can certainly be improved. Finally, just a few of them proposed solutions for the broad spectrum of problems that arise when implementing supervisory control in a PLC-based control system, as will be detailed later.

A contribution of the DECON9 methodology is that it allows dealing with various events at each PLC scan cycle, regardless if these events are controllable (can be disabled by control action) or not. Moreover, DECON9 provides a standardized approach and solution to many problems that arise while implementing SCT into PLCs.

The remaining of this chapter is organized as follows: In section 2, basic notations of the Supervisory Control Theory (SCT) for Discrete Event Systems (DES) control are introduced altogether with Monolithic and Local Modular approaches; Section 3 details the problems that arise while implementing SCT into a PLC; Section 4 presents the general assumptions behind the proposed methodology as well as its step-by-step detailed functioning; Section 5 presents a case study and how it can be solved by the methodology, and; Finally, section 6 concludes this chapter.

2. Supervisory control of discrete event systems

In the solution of manufacturing automation problems through the Supervisory Control Theory (SCT), the shop floor plant can be modelled as a Discrete Event System (DES) and finite-state automata are used to describe plant, specifications and supervisors. In this section, we introduce basic SCT notations. More details can be found in (Wonham, 2011) and in (Cassandras & Lafortune, 2008).

2.1 Discrete event systems

A Discrete Event System (DES) is a dynamic system that evolves in accordance with the abrupt occurrence of physical events at possibly unknown irregular intervals (Ramadge & Wonham, 1989). Application domains include manufacturing systems, traffic systems, software engineering, computer networks and communication systems, among others.

According to (Cassandras & Lafortune, 2008) and (Ramadge & Wonham, 1987, 1989) the free behaviour of a DES can be described through automata. An automaton can be represented by the 5-tuple $(Q, \Sigma, \delta, q_0, Q_m)$, where Q is the set of states, Σ is the alphabet of events, $\delta : Q \times \Sigma \to Q$ is the (partial) state transition function, q_0 is the initial state and $Q_m \subseteq Q$ is the set of marked states (Vieira et al., 2006). Σ^* is used to denote the set of all finite length sequences of events from Σ. A string (or trace) is an element of Σ^* and a language is a subset of Σ^*. A prefix of a string s is an initial subsequence of s, i.e. if r and s are strings in Σ^*, u is a prefix of s if $ur = s$. For a language L, the notation $\bar{L}$, called the prefix-closure of L, is the set of all prefixes of traces in L. L is said to be prefix-closed if $L = \bar{L}$ (Afzalian et al., 2010).

Consider that an automaton G represents the free behaviour of the physical system. Two languages can be associated with it: the closed behaviour $L(G)$ and the marked behaviour $L_m(G)$. The language $L(G)$ is the set of all sequence of events that can be generated by G, from the initial state to any state of G. Thus, $L(G)$ is prefix-closed because no event sequence in the plant can occur without its prefix occurring first. It is used to describe all possible behaviours of G. The language $L_m(G) \subseteq L(G)$ is the set of all sequence of events leading to marked states of G, each of them corresponding to a completed task of the physical system. A DES represented by G is said to be nonblocking if $\overline{L_m(G)} = L(G)$, i.e., if there is always a sequence of events which takes the plant from any reachable state to a marked state (Afzalian et al., 2010).

The concurrent behaviour of two or more DESs is captured by the synchronous composition of them. Thus, for two DES, G_1 and G_2, the synchronous composition is given by $G = G_1 \| G_2$. This expression can be generalized for any number of DES by $G = \|_{\forall i \in I} G_i$.

The automata can also be represented by transition graphs (see Figure 1), where the nodes represent the states and the arcs labelled with event names represent transitions. Usually, the initial state is identified by an ingoing arrow whereas a marked state is denoted by double circles.

2.2 Supervisory control theory

In the Ramadge & Wonham (1989) framework, the set of plant events Σ is partitioned into $\Sigma = \Sigma_c \,\dot{\cup}\, \Sigma_u$, two disjoint sets where Σ_c is the set of all controllable events and Σ_u is the set of all uncontrollable events. An event is considered to be controllable if its occurrence can be disabled by an external agent (named supervisor), otherwise it is considered uncontrollable. The necessary and sufficient conditions for the existence of supervisors are established in (Wonham, 2011).

A supervisor, denoted S, determines the set of events to be disabled upon each observed sequence of events. It is a map from the closed behaviour of G to a subset of events to be enabled $S : L(G) \to 2^\Sigma$. The controlled system is denoted by S/G (S controlling G) and is modelled by the automaton $G\|S$. The closed and the marked behaviour of the system under supervision are respectively represented by the following languages: $L(S/G) = L(S\|G)$ and $L_m(S/G) = L(S\|G) \cap L_m(G)$.

Further, S is said to be nonblocking if $L(S/G) = \overline{L_m(S/G)}$, i.e., if each generated trace of the controlled plant can be extended to be a marked trace of the controlled plant. Consider that a language $K \subseteq L_m(G)$ represents a control specification over the plant G. K is said to be controllable with respect to G (or simply controllable) if its prefix-closure $\bar{K}$ doesn't change under the occurrence of uncontrollable events in G. In other words, K is controllable if and only if $\overline{K}\Sigma_u \cap L(G) \subseteq \overline{K}$. Given a discrete event plant G and a desired nonempty specification language $K \subseteq L_m(G)$, there exists a nonblocking supervisor S such that $L_m(S/G) = K$ if and only if K is controllable with respect to G (Wonham, 2011).

However, if K is not controllable, the *supremal controllable sublanguage of K* with respect to G, denoted by $SupC(K,G)$, must be computed. In this case $L_m(S/G) = SupC(K,G)$ (Ramadge & Wonham, 1989).

In order to differentiate the controllability of events in the graph representation of automata, usually the state transitions due to controllable events are indicated by a short line drawn across the transitions (Chandra et al. 2003). Figure 1 represents an automaton, where the event A is controllable and the event B is uncontrollable.

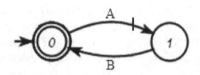

Fig. 1. A graph representation of an automaton

2.3 Monolithic approach

In the monolithic approach for the supervisors' synthesis (Ramadge & Wonham, 1989), the objective is to design a single supervisor that will coordinate the plant behaviour. Thus, all subsystems models G_i (where i is related to the number of subsystems), are composed in order to compute an automaton (generator) G that represents the free behaviour of the entire plant. In the same way, all control specifications E_j (where j is related to the number of control specifications) are composed into a global control specification E. From these models, one obtains the closed loop desired behaviour (known as target language) computing $K = G\|E$ and, consequently, obtaining a single supervisor S that marks the *supremal controllable sublanguage of K*, that is, $L_m(S) = L_m(S/G) = SupC(K,G)$.

According to (Queiroz & Cury, 2000a), in the monolithic approach the number of states of G grows exponentially with the number of subsystems. So this approach has the following drawbacks: the amount of computational effort when performing asynchronous composition of several automata, and; the use of supervisors with too many states in control platforms (usually a PLC) may generate extensive programs, where understanding, validation and maintenance will therefore, become difficult. In some cases the size of the program can render them unfit to be used in certain platform, either because of the storage or processing capacity.

In order to resolve these difficulties, Queiroz & Cury (2000a) propose using the local modular approach, as introduced in the next subsection.

2.4 Local modular approach

This approach is an extension to the monolithic approach and explores both the modularity of the plant and of the control specifications. It allows determining rather than a single and usually large supervisor, many local supervisors whose joint action guarantees the attendance of all the control specifications. Figure 2 illustrates the structure of local modular supervisory control for two supervisors.

Fig. 2. Local modular supervisory control architecture (Queiroz & Cury, 2002)

In this approach, the physical system behaviour is modelled by a Product System (PS) representation (Ramadge & Wonham, 1989), *i.e.*, by a set of asynchronous automata $G_i = (Q_i, \Sigma_i, \delta_i, q_{0_i}, Q_{mi})$, with $i \in N = \{1, 2, ..., n\}$, all of them having disjoint alphabets Σ_i. In turn, each specification imposed by the designer is represented by an automaton Ej with an alphabet $\Sigma j \subseteq \Sigma$, $j \in \{1, ..., m\}$, where m is the number of specifications. For each specification Ej a local plant G_{locj} is obtained, which is computed by the synchronous composition of all subsystems that share some events with the associated specification.

After determining all local plants it should be calculated the so-called local specification, which consists of performing synchronous composition between a given specification with its own local plant, *i.e.*, $K_{loc_j} = E_j \| G_{loc_j}$. Thus, the supremal controllable local sublanguages of K_{loc_j}, denoted by $SupC(K_{loc_j}, G_{loc_j})$, can be computed. Finally, it is possible to perform the synthesis of a local supervisor S_{loc_j} for each specification defined in the project. If at least one local supervisor disables the occurrence of an event, then the occurrence of this event is disabled in G (Vieira et al., 2006). To ensure that the system under the joint action of local supervisors is nonblocking, it should be guaranteed that the supervisors are nonconflicting, what is verified when the $\left\| {}_{j=1}^{m} \overline{L_m(S_{loc_j}/G_{loc_j})} = \right\| {}_{j=1}^{m} \overline{L_m(S_{loc_j}/G_{loc_j})}$ test holds. According to (Queiroz & Cury, 2000b), this condition ensures that the joint action of local supervisors is equivalent to the action of a monolithic supervisor that addresses all specifications simultaneously. Some computational tools can be used to assist in the synthesis of supervisors. For each one of them the models of subsystems and control specifications should be introduced in order to obtain synthesized supervisors, automatically. Among these tools *IDES* (Rudie, 2006), *TCT* (Feng & Wonham, 2006) and *"Grail for Supervisory Control"* (Reiser et al., 2006) can be mentioned.

3. Supervisory control implementation

3.1 Problems

According to (Fabian & Hellgren, 1998), *"the supervisor implementation is basically a matter of making the PLC behave as a state machine"*. However, this is not trivial task and can lead to many problems (Fabian & Hellgren, 1998):

Causality: SCT assumes that all events are spontaneously generated by the plant and that supervisors should only disable events generated by the plant. However, controllable events on practical applications are not spontaneously generated by the physical plant, but as responses to given PLC commands. Thus, for implementation purposes, "who generates what?" must be answered.

Avalanche Effect: occurs when a change on the value on a given PLC input signal is registered as an event that makes the software jump over an arbitrary number of states within the same PLC scan cycle. This may occur particularly if a specific event is used to trigger many successive state transitions, thus producing an avalanche.

Simultaneity: Due to the cyclical nature of the PLC processing in which input signals readings are performed only at the beginning of each scan cycle, the occurrence of uncontrollable events from the plant is recognized by the PLC once there are changes in the input signals values. Therefore, if in between successive scan cycles two or more signals change, they will all be recognized as simultaneous uncontrollable events regardless of their exact timing. As a result, the PLC is unable to recognize the exact order of uncontrollable events that happen in between scan cycles.

Figure 3 shows an automaton that is sensitive to the sequence of *B* and *C* events. Notice that depending on the order *B* and *C* events happens, the control decision is different, which highlights the problem that if *B* and *C* are recognized altogether in the same scan cycle, we would not be able to determine the actual state and, what should be the control action: *E* or *F*.

In order to avoid the simultaneity problem, the system must present the "interleave insensitivity" property (Fabian & Hellgren, 1998), which requires that after any interleaved uncontrollable events the "control decision" must necessarily be the same.

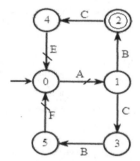

Fig. 3. Automaton that attempts to distinguish between the interleaving of events

Choice Problem: the supervisors obtained by the SCT are required to be "minimally restrictive", which means that the supervisors might provide alternative paths for the plant to choose from. Often a supervisor presents more than one possible controllable event from a single state. Thus, before producing a signal-change in the PLC outputs it may have to choose only one among them because according to Fabian & Hellgren (1998), generating more than one controllable event in a scan cycle can be contradictory and catastrophic.

Inexact Synchronization: during the program execution a change in any PLC input signal may occur and, this change will only be recognized at the beginning of the next scan cycle. The control reasoning is always performed on old frozen data. Therefore the communication between the PLC and the plant is subject to delays due to periodic reading of the input signals (Balemi, 1992). This inexact synchronization (Fabian & Hellgren, 1998) can be a problem when a change in a PLC input signal invalidates a control action (the choice made by the program, which corresponds to the generation of a controllable event).

3.2 Related work

Many researches have dealt with producing PCL programs from TCS. Some attempts (Fabian & Hellgren, 1998) did not propose a methodology but focused on solving particular situations which is far from a generic approach such as (Hasdemir et al., 2008). In the following, we briefly discuss some of these proposals.

In order to solve the choice problem, the solutions adopted in the literature follow the idea that for practical PLC implementation purposes, a deterministic controller must be statically extracted from the supervisor. Fabian & Hellgren (1998) also show that if a choice is not taken, the sequential execution of the program within the PLC will choose and the chosen transition will always be the same in a particular state according to the ordering of the rungs.

Moreover, Malik (2002) shows that depending on the choice taken (or deterministic controller extraction) the controlled behaviour may be blocking, even when the supervisor is nonblocking, which in that work is named *determinism problem*. In (Dietrich et al., 2002) three properties are given which, when satisfied, ensure that any controller for the system is necessarily nonblocking. In (Malik, 2002) a more general property is introduced and an algorithm is given which checks whether every deterministic controller generated from a given model is nonblocking. However, no controller can be constructed from those works in case the DES model does not satisfy these conditions. In particular, a valid controller may exist, even if the conditions of (Malik, 2002) and (Dietrich et al., 2002) do not hold. But, in (Morgenstern & Schneider, 2007) another approach to generate deterministic controllers from supervisors is presented and a property named *forceable nonblocking* is introduced.

In (Queiroz & Cury, 2002), the authors introduce a general control system structure based on a three level hierarchy that executes the modular supervisors' concurrent action and interfaces the theoretical model with the real system. They also propose a ladder-based PLC implementation of TCS but do not discuss the above mentioned problems.

In (Hasdemir et al., 2004) the authors propose the use of two bits for each state in order to solve the avalanche effect, but none of the other problems are discussed. In addition, only a single event is processed per supervisor at each PLC scan cycle.

Vieira (2007) presents a methodology that considers some of the problems but the program is structured as Sequential Function Charts (SFC), which is not so widespread among PLC programmers so far. Also, this methodology requires to change the automaton model in order to remove self-loops and there is no solution to the choice problem as well.

Most of the proposals found in the literature (Leduc, 1996; Hellgren et al., 2002; Queiroz & Cury, 2002) implements SCT in the ladder language. They have the same drawback: they deal with one single event per scan cycle. Thus, if between two scan cycles "n" changes occur in the PLC inputs, the program will take "n" scan cycles to deal with them. The best proposals so far handle one event per supervisor at each PLC scan cycle. This constraint help ensure existing approaches to deal with the avalanche effect and the choice problem (*determinism problem*). However, in this way the supervisor's update rate and actions will be lower than that obtained via traditional solution, without the use of the SCT.

The related work presented above shows that a ladder-based PLC complete methodology that solves recurring problems on implementing TCS supervisory control is still missing. Below we present a methodology that fills this gap.

4. DECON9 methodology

This section presents DECON9 (which comes from the main idea: DEcomposing the CONtrol Implementation DEpending on the CONtrolability of the events), a nine steps ladder-based PLC implementation of SCT supervisory control methodology that treats multiple events in the same scan cycle and also solves the avalanche effect and the choice problem. The methodology was inspired by the work of Queiroz & Cury (2002) but the Product System (PS, the asynchronous sub-plant models) and supervisors' implementations are decomposed into blocks of events according to their controllability.

At the beginning of each PLC scan cycle, all signal changes in the PLC inputs are registered as uncontrollable events in the PS level, and state transitions due these events are processed in the PS automata. Immediately after that, the supervisors also perform the state transitions due uncontrollable events. In this way the treatment of uncontrollable events are prioritized, and PS and supervisors are maintained in synchronization with the plant.

From the current state of the supervisors all events that are still disabled are verified through a disabling map. Thus, if at least one local supervisor disables a certain event, then the occurrence of this event is globally disabled. Thereafter, from the list of non-disabled events the choice problem (determinism problem) is inferred. If there is more than one enabled event in a current local supervisor state an event is randomly selected. In opposition to the other proposals in which a deterministic controller is statically extracted out of the supervisor (offline procedure) before being implemented, in our methodology the supervisors are fully implemented and the decisions on which controllable event may be executed are dynamically performed on the fly (during the program execution). So all alternative paths in the supervisor are preserved and the system behaviour under supervision is ensured to be nonblocking.

All enabled controllable events which are likely to occur in the plant are generated in the PS and the states are updated in the subsystems and supervisors models. Finally, these events are mapped to PLC outputs and another scan cycle begins.

Notice that the coherence of control actions is guaranteed because before defining the set of disabled events and generating the controllable events in the PS, the states of the subsystems and supervisors are all updated (this means that the supervisors know in which state they are in and which events are enabled).

4.1 Solving the avalanche effect: Damming

To avoid the avalanche effect the methodology indicates to use two auxiliary memories for each uncontrollable event: the first one is used to store the events generated by the plant and; the second is used to enable PS and supervisors to transit states. Every time the second memory is used, it is reset (deleted).

In this way, an event is used only once and multiple transitions are hold up. In any case the initial state of the event is required, the first memory can be used. As long as the PS is composed of asynchronous subsystems, once it is updated due to a given uncontrollable event, this information is not used any more until this same event is generated by the plant.

However, the information that a given uncontrollable event is active can be used somewhere in de program, especially to update supervisors states, once an uncontrollable event can be part of many supervisors simultaneously. Therefore, it should be possible to recover all information regarding uncontrollable events generated by the plant before updating supervisors. That's the reason for the second auxiliary memory.

4.2 Solving the choice problem: Random choice

To deal with the choice problem, one should analyse each supervisor at a time, to look after states where the problem may occur and which events are involved in each case. The simpler case is when there are only two controllable events to choose from, but situations involving a handful of choices are not rare in real applications. Figure 4 presents a flowchart for dealing with the choice problem. It helps identifying corresponding states and events.

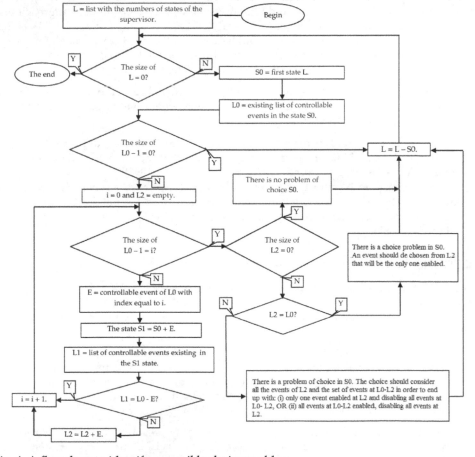

Fig. 4. A flowchart to identify a possible choice problem

To solve the choice problem, states that present multiple control alternatives need to be identified. But, if just after the disabling routine more than one controllable event is enabled at an active state, a routine is called that performs a random selection between pairs of these events. Randomness happens because an auxiliary memory is used to help perform the selection. This memory changes state at every scan cycle and, because there is no deterministic way to predict the number of scan cycles before the choice problem occurs, it acts as if it is random.

4.3 Solving simultaneity and inexact synchronization: Hardware interruption

For the simultaneity and inexact synchronization problems the solution adopted is the use of hardware interruption. Then, the uncontrollable events that may cause these problems must be associated with that type of PLC input. Thus, when a change in one of these PLC inputs occurs, the program is interrupted and the event is registered at the moment of its occurrence. It is important to notice that many PLCs do not have this kind of input and, in that case, there is no particular solution in DECON9.

It should be pointed out that these are not problems that arise exclusively while implementing SCTs into PLCs; they could happen in any given conventional approach that did not even use SCTs. Nevertheless, it can be said that an advantage of using SCTs is that these problems can be identified and we could be aware of them at the very beginning stage of designing the control system. On conventional approaches however, they can only be identified, if ever, after an extensive trial-and-error validation experiments.

On the other hand, not all systems' models will present this kind of problem. Thus, for the local modular supervisory control structure to be implemented without such problems the model must abide to some properties:

- To be sure that no problem regarding the "inability to identify event's order" problem will happen, it should be ensured that all automata that model every supervisor show the "interleave insensitivity" property (Fabian & Hellgren, 1998);
- Complementary, to be sure that no problem regarding "inexact synchronization" will happen, it should be ensured that the resulting language from every supervisor's automata, and their corresponding supervisors, show the "delay insensibility" (Balemi & Brunner, 1992).

5. DECON9 methodology step-by-step

This section will detail all nine steps of DECON9 methodology. DECON9 organizes the resulting program as a set of subroutine calls for the sake of better understanding, code reuse and reduction and, easy maintenance. Subroutine calls is a common feature available in almost every PLC.

Ten subroutines are created to fulfil all steps of DECON9 and they must follow a specific order. Figure 5 presents a complete flowchart where one can see all subroutine calls on the left and all nine steps on the right. It should be noticed that the third and fourth steps deal with uncontrollable events and the fifth to eighth steps deal with controllable ones. Also, "Make $Mx.0 = Mx.1$" routine is called twice for every scan cycle.

First step is to be performed at the very first scan cycle only because it sets initial states of all auxiliary memories that store the initial state of all supervisors and PS subsystems. The remaining states are reset.

Second step reads all PLC inputs and identify events coming from the plant according to signal changes that corresponds to uncontrollable events.

Third step promotes the state transitions for the whole PS altogether with all just identified uncontrollable events. For this end, only transitions belonging to PS that involves uncontrollable events are transited. At this point it should be reminded that each event transition performed produces the information on the event to be erased in order to avoid the avalanche effect.

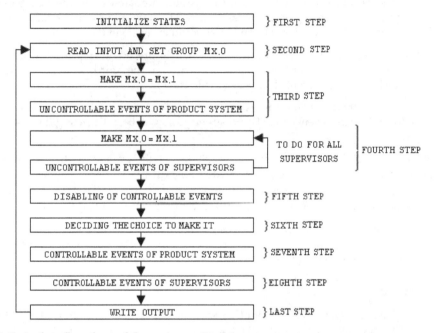

Fig. 5. Complete flowchart of the main routine

During the **fourth** step all supervisors must perform their state transitions considering uncontrollable events (and only these). The structure of the PLC program to be implemented for supervisors is the same as previously used for the PS system.

Because the information that a given event was active was erased during the transitions of the PS system, before updating the supervisors, the information of which uncontrollable events were generated by the plant must be recovered. For this end, all uncontrollable events use a pair of auxiliary memories: one of them to store which events were generated by the plant and the other is used for state transition and is discarded immediately after.

As a consequence, the methodology gives priority to uncontrollable events but do not neglect the synchronization of PS and supervisor states with the physical plant, therefore avoiding the avalanche effect.

The **fifth** step do not differ from what was proposed by Queiroz & Cury (2002) where, from the current state of each supervisor, the events disabled by any supervisor is disabled by the whole set of supervisors.

Sixth step starts off taking into account the status of all supervisors and a list of all still enabled controllable events. From these, if any supervisor shows the choice problem, it is resolved at this step. The program structure to solve this problem depends on the number of choices available (as presented earlier) but if no supervisor presents this problem, the sixth step does not produce any code.

As for the **seventh** step, it relates to the generation of controllable events that were not disabled beforehand and could possibly occur on the plant. This event generation is done at the PS level and is followed by the PS state transitions update due to the just generated events. Therefore, step seven is responsible for all state transitions related to PS's controllable events and completing the implementation of the PS in the PLC.

Eighth step updates all controllable events generated in preceding step in all supervisors. Therefore, the remaining transitions not dealt with at the third step are done here completing the implementation of all supervisors in the PLC.

Thus, as a result of the last two steps, even before a physical control at the PLC output is issued due to controllable events, the PS and supervisors will be anticipating the state of the physical plant.

It should be noticed that there is also the possibility of the avalanche effect problem for controllable events. However, DECON9 establishes no particular procedure to deal with them because it is understood that in a practical application this problem will not occur. In any case, if it ever happens, it can be treated the same way it was done for uncontrollable events.

Ninth, and last step, sends controllable events generated by the control logic to the physical plant. This is done by mapping events to specific drives at the PLC outputs. This action generates new events from the physical plant and another scan cycle begins.

6. Case study

In order to illustrate DECON9, a complete solution for supervisory control problem is presented. The case study covers the plant and specifications modelling, the synthesis of supervisors, up to the coding of the control logic in ladder, ready to be implemented in a PLC.

6.1 Description of the physical system

The problem to be studied consists of a transfer line with six industrial machines M_X (where $X = 1,..., 6$) connected by four buffers B_Y (where $Y = A, B, C, D$), capable of storing only one part, as shown in Figure 6. This problem was studied by Queiroz & Cury (2000a) and was chosen because it produces simple automata, is easy to understand, is fairly possible to occur as part of bigger layouts and presents some of the problems DECON9 deals with.

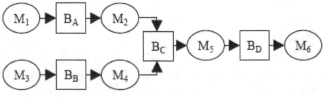

Fig. 6. Industrial transfer line case study

Start operation of the machines are controllable events, and the end operation are uncontrollable events. The transfer line should work in order to transport the parts through the machines, but a machine can't start operation if there is no part in its input buffer. Since the machines M_1 and M_3 have no input buffer, it should be considered that there will always be parts available for these machines. Similarly, M_6 can release as many parts as it is capable of producing.

6.2 Plant modelling

Table 1 presents a list of events associated with the operation of each machine as well as the type of event according to its controllability, the description of the event and, which PLC input (I) or output (Q) it is associated with.

DEVICE	EVENT	EVENT TYPE	DESCRIPTION	I/O
Machine 1	A_1	Controllable	Machine 1 start operation	Q0.0
	B_1	Uncontrollable	Machine 1 stop operation	I0.0
Machine 2	A_2	Controllable	Machine 2 start operation	Q0.1
	B_2	Uncontrollable	Machine 2 stop operation	I0.1
Machine 3	A_3	Controllable	Machine 3 start operation	Q0.2
	B_3	Uncontrollable	Machine 3 stop operation	I0.2
Machine 4	A_4	Controllable	Machine 4 start operation	Q0.3
	B_4	Uncontrollable	Machine 4 stop operation	I0.3
Machine 5	A_5	Controllable	Machine 5 start operation	Q0.4
	B_5	Uncontrollable	Machine 5 stop operation	I0.4
Machine 6	A_6	Controllable	Machine 6 start operation	Q0.5
	B_6	Uncontrollable	Machine 6 stop operation	I0.5

Table 1. Machine-related events for the case study

The behaviour of each M_x (where x = 1, .., 6) machine is represented by a G_x automaton shown in Figure 7. Each machine has only two states: in the state 0 the machine is stopped, waiting to work, and the state 1 the machine is working on a part. According to Table 1, the start operation is a controllable event A_x, and the stop operation is uncontrollable event B_x.

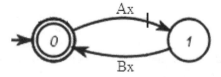

Fig. 7. G_X automaton for each machine

It is important to observe that passive devices need not be modelled, i.e., devices that don't have proper events, such as the buffers in the transfer line, for instance.

6.3 Control specifications modelling

Control specifications are models that describe the desired behaviour for the closed loop system.

The automaton presented at the left-side of Figure 8, shows the control specification of the buffers to avoid their overflow and underflow. It represents the working specification of B_A if $x = 1$, B_B if $x = 3$ and, B_D if $x = 5$. For all buffers, state 0 represents an empty buffer while state 1 represents a full buffer. The specification represented by the E automaton at the right-side of Figure 8 prevents the B_C buffer overflow and underflow. B_C will be full (state 1) if either B_2 or B_4 events occur and will be emptied (state 0) if an A_5 event happen. Therefore, machine M_5 will only be able to start operation after machine M_2 or M_4 produce a part in their output buffers. Once a part is deposited on a buffer, another part can only be deposited after a subsequent machine start operation (which signals that it collected a part from the buffer). Note that randomness must be guaranteed to prevent the machine M_5 from working with parts coming from only one of M_2 or M_4 machines.

Fig. 8. "E" specifications for B_A, B_B and B_D, buffers (on the left) and B_C (on the right)

6.4 Synthesis of local modular supervisors

From the devices (G_x) and operating specifications (E_Y) models, a synchronous composition between these models must be performed (as required by Queiroz & Cury, 2000b).

Firstly, you must determine the Product System (PS). To do this, the synchronous composition of all sub-plants that present common events should be performed. It should be looked for the biggest amount of asynchronous subsystems possible. For the present case study no common events between any sub-plants exist, therefore the models previously presented are the set of subsystems of PS.

Then the set of local plants must be determined. To do this, a synchronous product between the subsystems that are affected directly or indirectly by a particular specification must be done.

The most practical way to identify common events is through a table, like Table 2, so the local plants (those that share specifications) are: $G_{locA} = G_1 \| G_2$, $G_{locB} = G_3 \| G_4$, $G_{locC} = G_2 \| G_4 \| G_5$ and $G_{locD} = G_5 \| G_6$. From the common events between specifications and sub-plants analysis, it should be verified if some specification can be grouped together. For the present case study this grouping does not occur.

Following, a synchronous composition of local plants with the specifications should be performed to generate local specifications: $K_{locA} = G_{locA} \| E_A$, $K_{locB} = G_{locB} \| E_B$, $K_{locC} = G_{locC} \| E_C$, and $K_{locD} = G_{locD} \| E_D$.

Finally, the maximum controllable sublanguage of K_{locY} is calculate, which is denoted $SupC(K_{locY} \| G_{locY})$, where $Y = \{A, B, C, D\}$. The results are the local supervisors: S_{locA}, S_{locB}, S_{locC} and S_{locD}, that are presented in Figure 9 where the left side shows the supervisors S_{locA} if $z=1$, S_{locB} if $z=3$ and S_{locD} if $z=5$ and; at the right side the S_{locC} supervisor is shown.

	M_1		M_2		M_3		M_4		M_5		M_6	
	A_1	B_1	A_2	B_2	A_3	B_3	A_4	B_4	A_5	B_5	A_6	B_6
E_A	X	X										
E_B					X	X						
E_C				X			X	X				
E_D									X	X		

Table 2. Common events between sub-plants and specifications

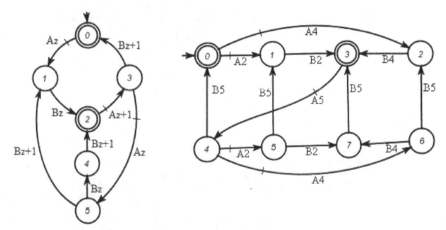

Fig. 9. Local modular supervisors

However, it is necessary to ensure the local modularity of local supervisors altogether so the joint action of all supervisors is nonblocking, as demonstrated by Queiroz & Cury (2000b). To verify local modularity, the synchronous composition of all local supervisors must be performed, as follows: $S = S_{locA} \| S_{locB} \| S_{locC} \| S_{locD}$. The resulting automaton from this composition should be checked for blocking states. If no blocking states can be found, then it can be said that local supervisors are modular to each other.

In order to reduce the amount of memory used in the implementation of these supervisors some tools to reduce the supervisors, these tools keep the control action that disable controllable events, but the supervisors lose information about the plant (Su & Wonham, 2004). However, as the product system will be implemented together with supervisors in the PLC, the information that was lost by reducing the supervisors will be preserved in the product system. Figure 10 shows the same supervisors of Figure 9, but in reduced form where the left-side show the S_{locA} (if $z=1$), S_{locB} (if $z=3$) and S_{locD} (if $z=5$) supervisors, and the S_{locC} supervisor is presented at the right-side.

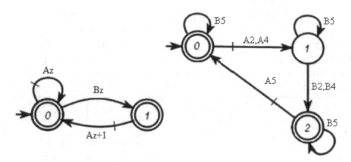

Fig. 10. Reduced supervisors automata

6.5 Following DECON9´s methodology

a.   Main Routine

To easy understand, the PLC program is organized as a main routine that calls subroutines for every single block in the flowchart of Figure 5. Figure 11 shows DECON9's main routine. The sequence of calls should be followed in such a way that this main routine works like a template for all systems. Therefore, there will be 10 (ten) subroutines that will be detailed following. Notice that some abbreviations were considered in order to simplify the PLC code. Thus, S_C is the abbreviation for S_{locC} and Sc.0 means state 0 of Supervisor S_C. Moreover, dAx is used to indicate the disabling of the Ax event. Thus, in Figure 11 dA2 and dA4 are used to indicate the disabling of A2 and A4, respectively.

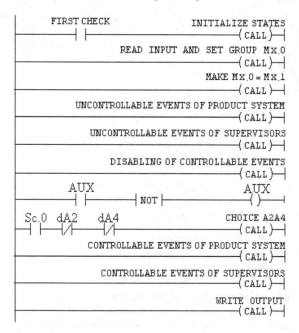

Fig. 11. DECON9 main routine

Figure 12 shows the subroutines in the order they will be called by the main routine. In order to facilitate understanding, the code for each subroutine is shown just below the line that promotes the corresponding call. Each of the subroutines is presented in sequence.

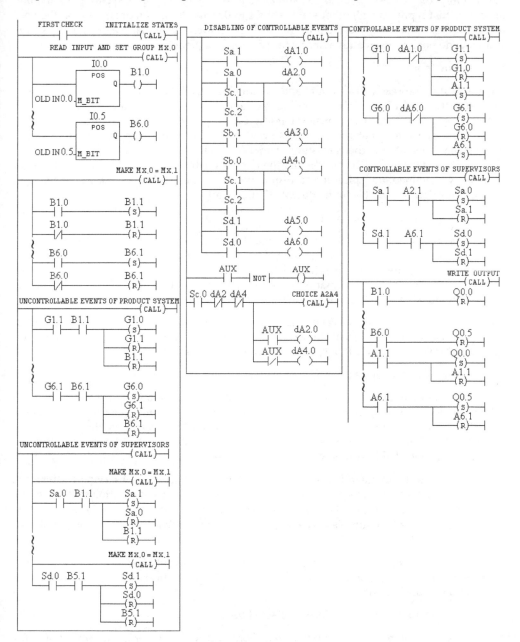

Fig. 12. PLC implementation for the case study

b.   State Initialization

The first subroutine initializes all states of the Product System (PS) and supervisors. Thus, the memory that corresponds to the initial state of all automata is set to 1. Remaining memories that represents all other states are set to 0. This should be done only on the very first scan cycle (that's why a memory flag called "first check" is used alongside it) so the automata do not lose its evolutionary feature during a sequence of scan cycles.

c.   Reading Inputs

Second subroutine reads PLC inputs and identifies controllable and uncontrollable events that came from the plant. This subroutine is called at the beginning of every scan cycle to verify if there is any positive transition at any PLC input. If so, there is an uncontrollable event being generated that corresponds to that input. The correspondence of inputs and uncontrollable events for the case study is in Table 1. It should be noticed that the "POS" PLC function (see Figure 12) ensures that the uncontrollable event will be identified only at the scan cycle immediately after the corresponding input signal changes (positive edge) and that this function is available to the RockWell PLC family.

d.   Rescuing Uncontrollable Events

Every uncontrollable event uses two memories, $Mx.0$ and $Mx.1$. The first group of memories, $Mx.0$, is responsible for storing the information of all events that actually have been produced by the plant. Therefore, there is a subroutine that updates the second set of memories ($Mx.1$) with the information stored at the other set ($Mx.0$) so the second group is used to promote the state transitions at PS as well as at supervisors.

As long as the information of each event that have been issued by the plant is stored in $Mx.0$, $Mx.1$ can eventually lose its information because it can always be recovered from $Mx.0$ by calling "Make $Mx.0 = Mx.1$" subroutine, as shown in Figure 12.

e.   Updating Product System with Uncontrollable Events

Next subroutine deals with PS uncontrollable events and is responsible for performing PS state transitions due to these events. It can be interpreted as an "automata player". There is no restriction on the number of events issued by the plant that this automata player is able to deal with. Therefore, at each scan cycle, PS automata can transit states regardless the number of uncontrollable events coming from the plant.

The current state of all subsystems is updated. This can be seen by observing for instance that when G_1 sub-plant is in state 1 and B_1 event happens, the state transition to 0 will occur and, to avoid the avalanche effect, $B1.1$ memory is reset to 0 (see Figure 12). It should be noticed that if any other sub-plant is able to promote state transitions, it will be possible to promote it as well.

Because PS is composed of asynchronous subsystem, an event that is dealt with in one subsystem will not occur in another. Thus, there is no problem of erasing its information when its state transition occurs.

f.   Updating Supervisors with Uncontrollable Events

Another automata player is implemented here but only to promote transitions for uncontrollable events of the supervisors. For these supervisors, that are not necessarily asynchronous, the same event can produce state transitions in more than one supervisor. Therefore, once the information on every event that promotes a transition is erased to avoid the avalanche effect, this information should be recovered before executing the automata player for each supervisor. That's why "Make $Mx.0 = Mx.1$" subroutine is called once before each supervisor in the case study, as illustrated in Figure 12.

Note that the program structure used to update the supervisors is the same used for the PS but the supervisors' states are considered instead.

g.   Disabling Controllable Events

According to the current state of all supervisors, all controllable events that should be disabled are determined.

Once PS and supervisors states are updated with the transitions promoted by the uncontrollable states issued by the plant, it can be said that all automata implemented into the PLC are in synchrony with the plant, i. e., they are all at the same states as the physical plant.

Therefore, it is possible to identify events that need to be disabled by the conjunction of the supervisors. It is possible that a single event became disabled by the action of many supervisors.

h.   The Choice Problem

This subroutine should be called only if necessary and, depending on the state of a given supervisor and on the events involved in the choice problem. For each choice problem that appears, a different subroutine must be created to deal with it.

It is possible that two or more controllable events became disabled by the supervisors. If they belong to the same supervisor a choice problem may occur. In the case study at hand, M_2 and M_4 machines cannot start operation at the same time because they share the same output buffer and thus, once one machine issue a part to that buffer, the other cannot issue another one. In other words, when supervisor S_C is in state 0, A_2 e A_4 events are enabled but cannot be issued at the same time (neither at the same scan cycle) because issuing one means disabling the other. This is a clear choice problem whereby the Product System must decide which one to issue.

According to the flowchart shown in Figure 4 that ensures a solution to the choice problem at the same time that it avoids rendering a blocking system, a "Choice A2A4" subroutine is called (see Figure 12). This subroutine randomly enables only one event at a time, either A_2 or A_4 for the supervisor S_C when it is in its 0 state, for the present case study. An auxiliary memory, called "AUX", is used which changes its state (from 0 to 1, and vice-versa) at every scan cycle. Therefore, when AUX holds 1, A_2 event is disabled and, when AUX holds 0, A_4 is disabled.

i.   Issuing Controllable Events from PS

Another automata player is implemented but only controllable events of the Product System are dealt with. Thus, each controllable event that has not been disabled and ready to occur would make PS to transit states and an event to be issued from PLC which means that a controllable event occur at the physical plant.

It is important to observe that the choice problem happens among non-disabled controllable events at a particular state of some supervisor and not among events of different supervisors. Thus, DECON9 allow that many controllable events can be issued at the same scan cycle. For instance, at the present case study, M_1 and M_3 start operation can happen at the same time and, as a consequence, A_1 e A_3 events can also be issued at the same scan cycle.

It should be noticed that every state transition that occur at PS corresponds to signalling a particular event that must be issued.

j.   Issuing Controllable Events from Supervisors

As can be noticed in Figure 12, controllable events of PS might promote state transitions on the supervisors. Therefore, another automata player is implemented here but only for supervisors' controllable events.

k.   Writing Outputs

Finally, at the end of the scan cycle, PLC outputs are updated. It should be noticed that all output reset conditions are implemented first and just afterwards, output signals are issued according to controllable events.

7. Conclusions

Supervisory Control Theory (SCT) of Discrete Event Systems (DES) has become a major player in the next step manufacturing system automation once it brings formality, predictability and a higher abstract level of specification to the analysis of complex layouts. Some of the advantages of using SCT include: plant and supervisors are high level models; testing of resulting control program is not required once it is produced from a sound theoretical background; equipment or plant behaviour models can be easily reused and; better control programs can be achieved by engineers focusing on the modelling instead of the intricacies of implementing it.

But widespread use of SCT has been hold up by the fact that Programmable Logic Controllers (PLCs) are the basic devices that can be found in the shop-floor. Implementing SCT in PLCs is not a trivial task because many problems and constraints arise while attempting to do it. Many researches have dealt with producing PCL programs from TCS. Some attempts did not propose a methodology but focused on solving particular situations which is far from a generic approach. Most of the existing proposals are based on the monolithic approach for the supervisors' synthesis, and the implementation is performed in ladder language. In some works the synthesis of supervisors is performed according to the local modular approach, which reduces the computational complexity of the synthesis process and the size of supervisors by exploiting specifications modularity and the decentralized structure of composite plants. But almost all of them can only tackle one event per PLC scan cycle, which represents a problem when handling large scale plants. Moreover, in this way the supervisor's update rate and actions will be lower than that obtained via traditional solution, without the use of the SCT. Finally, just a few of them proposed solutions for the broad spectrum of problems that arise when implementing supervisory control in a PLC-based control system.

In this chapter a nine step methodology, named DECON9, was presented. DECON9 is a methodology to implement SCT into PLCs in standardized, efficient and robust ways, closer to real size plants. It is a standardized approach because represents a complete methodology for the whole process, and is divided into simple sub-routines. It can deal with large scale plants because it uses the local modular approach for the supervisors' synthesis. It is an efficient solution because can tackle more than one event per scan cycle. It is robust because can predict problems and solves some of the most common ones. It turns PLCs into a state-machine where supervisors and plant events are explicitly represented and their control reasoning depends on their controllability.

This chapter also reviewed the basics of DES and the problems of implementing SCT into a PLC, presented detailed functioning and implementation of DECON9 and gave an example on how to apply it.

The local modular approach was used for the synthesis of supervisors and their implementation in PLC was programmed using the well-known ladder language. DECON9 use is exemplified by the implementation of the supervisory control of an industrial transfer line case study found in the literature. Using this case study, it is demonstrated that DECON9's advantages include: (*i*) it allows the control logic to deal with many events at each scan cycle, which improves existing approaches that are constrained to only one event at a time; (*ii*) a nonblocking property is achieved thanks to the random selection of controllable events approach that solves the choice problem; (*iii*) there is no fear of an avalanche effect thanks to the use of auxiliary memories, and; (*iv*) uncontrollable events are prioritized.

A computational tool for automatic generation of PLC programs obtained through the Supervisory Control Theory (SCT) is under development. This tool will comply with DECON9. With this tool, the gap between theory and practice will be reduced even further thanks to the automatic procedure based on a sound methodology.

8. Acknowledgment

The authors would like to thank *Pollux Automation Company* and *Santa Catarina State University (UDESC)* for their support to pursue this work.

9. References

Afzalian, A.; Noorbakhsh, M. & Navabi, A. (2008). PLC implementation of decentralized supervisory control for dynamic flow controller. *Proceedings of the 17th IEEE International Conference on Control Applications (CCA'08)*, pp. 522-527, San Antonio, Texas (USA), September, 2008.

Afzalian, A. A.; Noorbakhsh, S. M. & Wonham, W. M. (2010). Discrete-Event Supervisory Control for Under-Load Tap-changing Transformers (ULTC): from synthesis to PLC implementation, In: *Discrete Event Simulations*, Aitor Goti (Ed.), pp. 285-310, InTech, ISBN 978-953-307-115-2, Retrieved from
<http://www.intechopen.com/download/pdf/pdfs_id/11550>

Ariñez, J.F.; Benhabib, B.; Smith, K.C. & Brandin, B.A. (1993). Design of a PLC-Based Supervisory-Control System for a Manufacturing Workcell, *The Canadian High Technology Show and Conference*, Toronto, 1993.

Balemi, S. (1992). *Control of Discrete Event Systems: Theory and Application*, Ph.D. thesis, Swiss Federal Institute of Technology, Zürich, Switzerland.

Balemi, S. & Brunner, U. A. (1992). Supervision of discrete event systems with communication delays, *Proceedings of the American Control Conference*, pp. 2794-2798, Chicago, IL, USA, June, 1992.

Cassandras, C. G. & Lafortune, S. (2008). *Introduction to Discrete Event Systems. (2nd edition)*, Springer, ISBN: 978-0-387-33332-8, New York, USA.

Chandra, V.; Huang, Z. & Kumar, R. (2003) Automated Control Synthesis for an Assembly Line Using Discrete Event System Control Theory. *IEEE Transactions on Systems, Man, and Cybernetics – Part C: Applications and Reviews*, Vol. 33, No. 2, (may 2003), pp. 284-289, ISSN 1094-6977.

Dietrich, P.; Malik, R.; Wonham, W. M. & Brandin, B. A. (2002). Implementation considerations in supervisory control. In: *Synthesis and Control of Discrete Event Systems*, Caillaud, B.; Darondeau, P.;Lavagno, L.; Xie, X. (Eds), pp. 185-201, Kluwer Academic Publishers.

Fabian, M. & Hellgren, A. (1998). PLC-based implementation of supervisory control for discrete systems, *Proceedings of the 37th IEEE Conference on Decision and Control*, Vol. 3, pp. 3305-3310.

Feng, L. & Wonham, W. M. (2006). TCT: a computation tool for supervisory control synthesis, *Proceedings of the 8th International Workshop on Discrete Event Systems – WODES*, pp. 388-389, Ann Arbor, Michigan, USA, July 2006.

Hasdemir T., Kurtulan, S., & Gören, L. (2004). Implementation of local modular supervisory control for a pneumatic system using PLC. *Proceedings of the 7th International Workshop on Discrete Event Systems (WODES'04)*, pp. 27-31, Reims, France.

Hasdemir, T.; Kurtulan, S.; & Gören, L. (2008). An Implementation Methodology for Supervisory Control Theory. *International Journal of Advanced Manufacturing Technology*, Vol. 36, No. 3, (March 2008), pp. 373-385. ISSN 0268-3768.

Hellgren, A.; Lennartson, B. & Fabian, M. (2002). Modelling and PLC-based implementation of modular supervisory control, *Proceedings of the 6th International Workshop on Discrete Event Systems (WODES'02)*, pp. 1-6. ISBN: 0-7695-1683-1. Zaragoza, Spain, October 2002.

Lauzon, S. C. (1995). *An implementation methodology for the supervisory control of flexible-manufacturing workcells*, M.A. Sc. Thesis, Mechanical Engineering Dpt. University of Toronto, Canada.

Lauzon, S. C.; Mills, J. K.; & Benhabib, B. (1997). An Implementation Methodology for the Supervisory Control of Flexible Manufacturing Workcells, *Journal of Manufacturing Systems*, Vol. 16, No. 2, pp. 91-101.

Leal, A. B.; Cruz, D.L.L.; Hounsell, M.S. (2009). Supervisory Control Implementation into Programmable Logic Controllers. In: *Proceedings of the 14th IEEE International Conference on Emerging Technologies and Factory Automation - ETFA*, pp. 899-905, Palma de Mallorca, 2009.

Leduc, R. J. (1996). *PLC Implementation of a DES supervisor for a manufacturing testbeb: an implementation perspective*, M.A.Sc. Thesis, Dept. of Electrical and Computer Engineering, Univ. of Toronto, Canada.

Leduc, R .J. & Wonham W. M. (1995). PLC implementation of a DES supervisor for a manufacturing test bed. *Proceeding of Thirty-Third Annual Allerton Conference on Communication, Control and Computing*, pp. 519-528, University of Illinois.

Liu, J. & Darabi, H. (2002). Ladder logic implementation of Ramadge-Wonham supervisory controller, *Proceedings of the 6th International Workshop on Discrete Event System (WODES'02)*, pp. 383-389. ISBN: 0-7695-1683-1. Zaragoza, Spain, October 2002.

Malik, P. (2002). Generating Controllers from Discrete-Event Models. In: F. Cassez, C. Jard, F. Laroussinie, M. D. Ryan (Eds.), *Proceedings of the Summer school in MOdelling and VErification of Parallel processes (MOVEP)*, pp. 337-342.

Manesis, S. & Akantziotis, K. (2005). Automated synthesis of ladder automation circuits based on state-diagrams. *Advances in Engineering Software*, 36, pp .225–233.

Morgenstern, A. & Schneider, K. (2007). Synthesizing Deterministic Controllers in Supervisory Control In: *Informatics in Control, Automation and Robotics II.* Filipe, J.; Ferrier, J-L.; Cetto, J.A.; Carvalho, M. (Eds), pp. 95-102, Springer, ISBN 978-1-4020-5626-0, Netherlands.

Noorbakhsh, M. & Afzalian, A. (2007a). Design and PLC Based Implementation of Supervisory Controller for Under-load Tap-Changer. *Proc. of the 2007 IEEE Int. Conf. on Control, Automation and Systems (ICCAS'07)*, pp. 901-906, Seoul, Korea.

Noorbakhsh, M. & Afzalian, A. (2007b). Implementation of supervisory control of DES using PLC. *15th Iranian Conf. on Electrical Engineering (ICEE'07)*, (in Farsi), Tehran, Iran.

Noorbakhsh, M. (2008). *DES Supervisory Control for Coordination of Under-Load Tap-Changing Transformer (ULTC) and a Static VAR Compensator (SVC)*. M.A.Sc Thesis, Dept. of Electrical & Computer. Eng., Shahid Abbaspour University of Technology, (in Farsi), Tehran, 2008.

Noorbakhsh, M. & Afzalian, A. (2009). Modeling and synthesis of DES supervisory control for coordinating ULTC and SVC. *Proceedings of the 2009 American Control Conference (ACC' 09)*, pp. 4759-4764, Saint Louis, Missouri USA, June 10-12, 2009.

Possan, M. C.; Leal, A. B. (2009). Implementation of Supervisory Control Systems Based on State Machines. In: *Proceedings of the 14th IEEE International Conference on Emerging Technologies and Factory Automation – ETFA*, pp. 819-826, Palma de Mallorca, 2009.

Queiroz, M. H. de & Cury, J. E. R. (2000a). Modular supervisory control of large scale discrete event systems, In: *Discrete Event Systems: Analysis and Control*. 1st Ed. Massachusetts: Kluwer Academic Publishers, pp. 103-110. Proc. WODES 2000.

Queiroz, M. H. de & Cury, J. E. R. (2000b). Modular control of composed systems. In: *Proc. of the American Control Conference*, pp. 4051-4055, Chicago, USA, 2000.

Queiroz, M. H. de & Cury, J. E. R. (2002). Synthesis and implementation of local modular supervisory control for a manufacturing cell, *Proceedings of the 6th International Workshop on Discrete Event Systems (WODES)*, pp. 1-6. ISBN: 0-7695-1683-1. Zaragoza, Spain, October 2002.

Qiu, R. G. & Joshi, S. B. (1996). Rapid prototyping of control software for automated manufacturing systems using supervisory control theory. ASME, *Manufacturing Engineering Division*, 4, pp. 95-101.

Ramadge, P. J. & Wonham, W. M. (1987). Supervisory control of a class of discrete event processes, *SIAM Journal on Control and Optimization*, Vol. 25, No. 1, pp. 206 - 230.

Ramadge, P.J. & Wonham, W.M. (1989). The control of discrete event systems, *Proceedings of the IEEE*, Vol. 77, No. 1, pp. 81-98.

Reiser, C.; Cunha, A. E. C. da & Cury, J. E. R. (2006). The Environment Grail for Supervisory Control of Discrete Event Systems, *Proceedings of the 8th International Workshop on Discrete Event Systems* – WODES, pp. 390-391, Ann Arbor, Michigan, USA, July 2006.

Rudie, K. (2006). The Integrated Discrete-Event Systems Tool, *Proceedings of the 8th International Workshop on Discrete Event Systems* – *WODES'06*, pp. 394-395, Ann Arbor, Michigan, USA, July 2006.

Silva, D. B.; Santos, E. A. P.; Vieira, A. D. & Paula, M. A. B. (2008). Application of the supervisory control theory in the project of a robot-centered, variable routed system controller, *Proceedings of the 13th IEEE International Conference on Emerging Technologies and Factory Automation* – *ETFA'08*, pp. 1-6.

Su, R. & Wonham, W. M. (2004). Supervisor reduction for discrete-event systems, *Discrete Event Dynamic Systems*, Vol. 14, No. 1, pp. 31-53.

Uzam, M.; Gelen, G. & Dalci, R. (2009). A new approach for the ladder logic implementation of Ramadge-Wonham supervisors, *Proceeding of the XXII International Symposium on Information, Communication and Automation Technologies (ICAT'09)*, pp. 1-7. ISBN: 978-1-4244-4220-1, Bosnia 2009.

Vieira, A. D.; Cury, J. E. R. & Queiroz, M. (2006). A Model for PLC Implementation of Supervisory Control of Discrete Event Systems. *Proceedings of the IEEE Conference on Emerging Technologies and Factory Automation - ETFA'06*, pp. 225-232. Czech Republic, September 2006.

Wonham, W. M. (2011). *Supervisory Control of Discrete-Event Systems*, The University of Toronto, available from: http://www.control.utoronto.ca/DES.

8

Integrated Cellular Manufacturing System Design and Layout Using Group Genetic Algorithms

Michael Mutingi[1] and Godfrey C. Onwubolu[2]

[1]*National University of Singapore, Electrical & Computer Engineering*
[2]*School of Applied Technology, HITAL, Toronto*
[1] *Singapore*
[2]*Canada*

1. Introduction

Cellular Manufacturing System (CMS), an application of group technology philosophy, is a recent technological innovation that can be used to improve both productivity and flexibility in modern manufacturing environments (Signh, 93; Sarker and Xu, 1998). In practice, the essence of CMS is to decompose a manufacturing system into manageable autonomous subsystems (called manufacturing cells) so as to enhance shop-floor control, material handling, tooling, and scheduling. The decomposition process involves identification of part families with similar processes or design features and machine cells so that each family can possibly be processed in a single cell. In addition to this, machine layout within each cell is considered essential in order to improve efficiency and effectiveness of the overall production system. Consequently, setup times, work-in-process inventories, and throughput times are reduced significantly. The overall process of designing CMS involves the following four generic phases:

1. *Cell formation*: involves grouping of machines which can operate on a product family with little or no inter-cell movement of the products.
2. *Group layout*: includes layout of machines within each cell (intra-cell layout), and layout of cells with respect to one another (inter-cell layout).
3. *Group scheduling*: involves scheduling of parts for production
4. *Resource allocation*: assignment of tools, manpower, materials, and other resources

In general, the design of CMS includes three critical decisions, namely, cell formation, group layout, and scheduling. In the most ideal case, these criteria should be addressed simultaneously so as to obtain the best possible results (Kaebernick and Bazargan-Lari, 1996; Mahdavi and Mahadevan, 2008). However, due to the complex nature of the decision problem coupled with the limitations of conventional approaches, most of the cell formation studies have focused on these decisions independently or sequentially (Selim, 1998; Onwubolu and Mutingi, 2001). Most cell formation approaches proposed in literature use flow patterns of parts (sequence data) for cell design issues only. On the other hand, the layout designers did not consider the cell formation problem. Due to the fact that the sequential approach addresses the cell formation and the cell layout problem in a disjointed fashion, the quality of the final solution is often compromised. In this chapter, an integrated approach to cell formation and layout design is presented, based on available sequence data. The GGA-based approach utilizes sequence data to identify machine cells as well as machine layout within each cell. In this view, the major objectives for this chapter are as follows:

- to develop a GGA based methodology for solving the integrated CMS design and layout problem using sequence data, or flow patterns.
- to develop relevant performance metrics to address the integrated cell formation and layout problem.
- to make a comparative analysis between the GGA approach and other well known algorithms found in literature.

The next section describes the cell formation and cell layout problem. Section 3 briefly explains the general GA framework. A GGA approach is presented in section 4. Section 5 provides the results and discussion. Finally, section 6 concludes this chapter.

2. Cell formation and layout problem

The cell formation problem (CFP) in CMS involves grouping of machines which can operate on a product family with similar manufacturing processes and features such that little or no inter-cell movement of products is involved. The overall objective of cell formation is to gain the advantages inherent in the philosophy of group technology. In assessing the quality of solutions, various objectives are considered. These objective functions include the following;

i.   Minimization of inter-cell movements;
ii.   Minimization of intra-cell movements;
iii. Maximization of utilization;
iv.   Minimization of material handling costs, and
v.   Minimizing cell work-load imbalances

The cell layout problem involves layout machine within each cell and layout of cells with respect to one another. Recently, researchers have made efforts to utilize interval data and ordinal data, consisting of process sequence data which identifies the order in which jobs are processed (Nair and Narendran, 1998; Won and Lee, 2001; Jayaswal and Adil, 2004). The application of sequence data in CMS has received little attention in the research community and in industry. Sequence data provides useful information on flow patterns of jobs in a manufacturing system. As such, sequence data is useful not only in identifying part family and machine groups but also the actual layout of machines within each cell, based on flow patterns. Earlier studies focused on the use of zero-one machine-component incidence matrix as the input data for the cell formation problem. However, the joint CFP and the layout problem are often treated independently in literature. In an attempt to jointly address the CF and the layout problem, solution methods from various researchers and practitioners often utilize a sequential approach. In this approach, cells are formed first, followed by intra-cell layout construction. Since the final solution is largely dependent on the initial cell formation, the quality of the final solution is often compromised.

The joint cell formation and layout problem is a new approach that seeks to identify manufacturing cells and the layout (sequence) of machines in the cells in an integrated manner. The whole aim of the approach is to avoid compromising the quality of solutions with respect to cell formation and cell layout objectives. Therefore, this approach to the joint layout problem is of practical value. The basic cell formation problem is NP-complete, meaning that it has no known polynomial time algorithm due to its combinatorial nature (Kumar *et al*, 1986). It follows that the integrated cell formation and cell layout problem is highly computationally intractable. In this respect, the use of heuristic approaches such as simulated annealing, tabu search, and genetic algorithms, is quite appropriate. Simulated annealing is a probabilistic meta-heuristic method proposed in Kirkpatrick, Gelatt and Vecchi (1983) and in Cern (1985) for finding the global minimum of a cost function that may possess several minima. It works by emulating the physical process whereby a solid is slowly cooled down so that when its structure eventually frozen, this occurs at a minimum energy configuration. Tabu Search is a meta-heuristic local search algorithm created in Glover and McMillan (1986) for solving combinatorial optimization problems. It uses the concept of a local or neighbourhood search to iteratively move from one potential solution x to an improved solution x' in the neighbourhood of x, until some stopping criterion has been satisfied, usually an attempt limit or a score threshold (Glover, 1989; Glover, 1990).

3. Genetic algorithms

Genetic algorithm (GA), originated by Holland (1975), is a meta-heuristic approach based on evolutionary principles of natural selection and survival of the fittest. The GA methodology has been applied extensively in a wide range of combinatorial problems in engineering, business, manufacturing, agriculture, telecommunications and sciences (Gen and Cheng; Goldberg, 1989; Man *et al*, 1999). The method integrates the elements of stochastic and direct search to obtain optimal (or near-optimal) solutions within reasonable computation time. GA attempts to evolve a population of candidate solutions by giving preference of survival to quality solutions, whilst allowing some low quality solutions to survive in order to maintain a level of diversity in the population. This process enables GA to provide good solutions so as to avoid premature convergence. Each candidate is coded into a string of digits, called chromosomes. New offspring are obtained from probabilistic operators, mainly crossover and mutation. Comparison of new and old (parent) candidates is done based on a given objective or fitness function so as to retain the best performing candidates into the next population. In this process, characteristics of candidate solutions are passed from generation to generation through probabilistic selection, crossover, and mutation actuated in the population of candidate solutions.

The general GA framework can be represented as follows:

```
BEGIN
        Initialize population with random candidate solutions;
        Evaluate each candidate;
REPEAT
        Select parent chromosomes;
        Recombine pairs of parents;
        Mutate the resulting offspring;
        Evaluate new candidates;
        Select individuals for next generation
UNTIL (Termination condition is satisfied)
```

Figure 1 shows the general flow of the genetic algorithm.

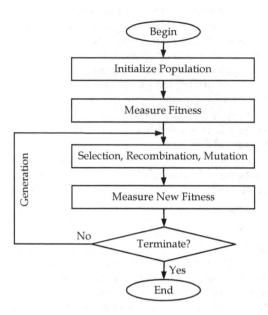

Fig. 1. Genetic algorithm framework

Genetic algorithm offers unique advantages over other stochastic searches, population-based search, including implicit parallelism, independence from gradient information, and flexibility to hybridization with other heuristics. Early applications of the GA approach to the cell formation problem include the work by Venugopal and Narendran (1992) based on minimization of cell load variation and inter-cell moves. Other applications were done by Gravel et al. (1998), and Hsu and Su (1998). However, Falkenauer (1992) realized several significant shortcomings of using classical GAs for grouping problems. Falkenauer (1998) pointed out that though attempts have been made to minimize the drawbacks associated with applying GAs to grouping problems by use of specialized genetic operators, this still result in various shortcomings. In this view, Falkenauer (1992) introduced a grouping genetic algorithm, designed to handle the special structure of grouping problems. Group genetic algorithm (GGA) is a modification of conventional GA designed specifically for clustering/grouping problems. In the next section, an enhanced GGA approach is proposed for the machine cell formation and layout problem.

4. A group genetic algorithm approach

Grouping genetic algorithm (GGA) combines specifically designed operators for grouping problems with the power of local search in order to refine new chromosomes generated. Therefore, GGA is a preferable approach over other heuristic and conventional approaches. The design of the proposed GGA for the joint cell design and layout problem is presented, based on its six main building blocks, namely:

i. Fitness/objective function
ii. Chromosome coding scheme
iii. Initial population generation
iv. Selection and recombination
v. Group genetic operators: crossover, mutation and inversion
vi. Genetic parameters

The next sections elaborate on these building blocks.

4.1 Objective/fitness function formulations

From the CMS design perspective, the existence of voids and exceptions should be minimized. In layout design, adjacency of machines in a cell is a key factor as it can reduce material handling costs significantly (Mahdavi and Mahadevan, 2008). From production planning perspective, the sequence in which machines are placed in cells may create unwanted reverse flows and skipping of workstations. For instance, a cell with machines 1 and 2 has two possible sequences (layouts), i.e., [1, 2] or [2, 1]. From Table 1, it can be seen that the cell layout [2, 1] has only one consecutive forward flow, while layout [1, 2] has four. From this analysis, layout [1, 2] is preferred.

		Parts									
	2	1	4	5	3	7	6	8	9	11	10
Machines											
2	2	2	3	2	2	2					
1	1	1	1	3	1	1					
5							1	1	1	2	1
3			2	1			3	2	2	3	2
4							2	3	3	1	3

Table 1. A typical solution for a cell formation problem

Ideally, a good objective function should be able to capture and evaluate the effects of the sequence of machines within each cell. A simplified way of evaluating the fitness of a cell layout is to express the objective function in terms of the number of consecutive forward flows. In this connection, Mahdavi and Mahadevan (2008) defined the cell flow index (CFI) and the overall flow index (OFI) for evaluating the performance of cell design and cell layout solutions.

The following notation is used in this model.

n number of parts in the system
m number of machines in the system
n_c number of parts in cell c
m_c number of machines in cell c
v_c number of voids in cell c
N_{fc} number of consecutive forward flows within cell c
S_{jk} machine-component matrix [s_{jk}]; $s_{jk} = 1$ if part k visits machine k, and 0 otherwise

In order to determine the average flow and overall flow performance measures, the total number of operations and the consecutive flows between a pair of machines are calculated. The total number of flows N_{flow} is:

$$N_{flow} = \sum_k \max_j s_{jk} - n \qquad (1)$$

The total number of flows in each cell c is determined as follows:

$$N_{tc} = (n_c m_c) - v_c - n_c \qquad (2)$$

4.1.1 Cell flow index (CFI)

The cell flow index for cell c, CFI_c is the ratio of the number of consecutive forward flows to the total number of flows within the cell. The cell average flow index is the weighted average of CFIs. This is further explained in the following expressions;

$$CFI_c = \frac{N_{fc}}{N_{tc}} \qquad (3)$$

Therefore, the average cell flow index, ACFI is

$$ACFI = \left(\frac{1}{n}\right) \cdot \sum_c n_c CFI_c \qquad (4)$$

It is clear from the above analysis that as the number of voids in the cell decreases and as the number of consecutive forward flows increases, the CFI measure increases. This indicates that the CFI represents the solution quality with respect to the number of voids and the intra-cell moves. Therefore, a combination of these performance measures ensures that the cell formation and layout are addressed jointly.

4.1.2 Overall cell flow index (OFI)

The OFI performance measure defines the ratio of the sum of consecutive forward flows in all the cells to the total number of the flows required to process all the parts. This can be expressed as follows;

$$OFI = \left(\frac{1}{N_{nflow}}\right) \cdot \sum_c N_{fc} \qquad (5)$$

Expression (5) shows that the overall cell flow index defines the extent of inter-cell moves (exceptions); increasing values of OFI can be obtained by decreasing values of inter-cell moves. While the OFI points to the inter-cell movements, the ACFI addresses the intra-cell movements.

4.2 Solution encoding – chromosome representation

The GGA's performance strongly depends on the type of the coding scheme, that is, the chromosome (string) representations used. Effective coding schemes can improve the search efficiency and quality. Most of the coding schemes in literature used strings of integer numbers to where the position of the number represents the machine and the value of the number identifies the cell number. For example, a typical chromosome (2 3 1 1 2 3 1 1 2) containing 9 machines represents a manufacturing system with 3 cells. Machines 1, 5, 6 and 8 are in cell 1, machines 2 and 3 are in cell 2, and machines 4, 7, 9 occupy cell 3.

Machine position:	1 2 3 4 5 6 7 8 9
Chromosome :	2 3 1 1 2 3 1 1 2

The proposed GGA algorithm has an improved coding scheme, similar to the one proposed Filho and Tibert (2006). The coding scheme improves the utilization of the group structure by using a group structure for each feasible string based on three code schemes as shown in Figure 2. The first, code 1, is a string of size m, where m represents the total number of machines in the system. The second is a group structure upon which the genetic operators act, while the third represents the positions of the last nodes of each group.

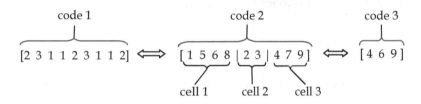

Fig. 2. Chromosome representation

It can be seen from code 3 in Figure 2 that cell 1 consists of the first four genes in code 2. Similarly, cell 2 is made up of the next two genes. Finally, cell 3 is composed of the last three genes in code2. Several features are enhanced in the implementation of the GGA structure, such as in formulation of objective/fitness functions, the genetic operators, chromosome repair and other genetic strategies.

4.3 Initial population

An initial population of the desired size, *popsize*, is randomly generated from the solution space. Consider a typical problem consisting of m machines and a predetermined number of cells, v. Assume that each cell comprises at least two machines. Then, the initial population is created according to the following procedure:

Repeat

1. For each cell j (j=1,...,v), randomly select two machines from the set of machines.
2. For the remaining (n-2j) unassigned machines, randomly assign a machine to a cell, until all machines are assigned.
3. Encode the chromosome using code 1 and add to the initial population.

Until (population size *popsize* is achieved).

In GGA application, the goal is to minimize some cost function which is usually mapped to a score function which is used to evaluate the generated chromosomes. A mapping procedure initially suggested by Goldberg (1989) is applied as follows;

$$f^i(t) = \begin{cases} f^i_{max} - g^i(t) & \text{if } g_i(t) < f^i_{max} \\ 0 & \text{if otherwise} \end{cases} \tag{6}$$

where, $g(t)$ is the objective function of a chromosome and f_{max} is the largest objective function in the current population.

4.4 Selection strategy

Several selection strategies have been suggested by Goldberg (1989), such as deterministic sampling, remainder stochastic sampling with/without replacement, stochastic tournament, and stochastic sampling with/without replacement. The remainder stochastic sampling without replacement has been found to be the most effective and is applied in this work (Goldberg, 1989). In this strategy, each chromosome i is selected and stored in the mating pool according to the expected count e_i calculated as,

$$e_i = \frac{f_i}{(1/popsize)\sum_{i=1}^{s} f_i} \tag{7}$$

Where, *popsize* is the desired population size and f_i is the score function value of the ith chromosome.

Each chromosome receives copies equal to the integer part of e_i, that is, $[e_i]$, while the fractional part is treated as success probability of obtaining additional copies of the same chromosome into the mating pool.

4.5 Genetic operators

In this section, design issues relating to the development of the proposed GGA approach for the manufacturing cell design problem are defined. Unique crossover, mutation and inversion strategies are developed for the GGA algorithm.

4.5.1 Crossover

Crossover is a probabilistic evolutionary mechanism which seeks to mate chromosomes, chosen by the selection strategy, in order to produce a pool of new offsprings, called *selection pool*. It allows the algorithm to generate new solutions and to explore unvisited regions in the solution space. The proposed crossover, called group crossover operator, exchanges groups of genes of selected chromosomes. The crossover operation occurs with probability *prcoss* until the desired pool size, *poolsize = popsize · pcross* , is obtained. The procedure for the group crossover operator is as follows:

Repeat

1. Generate a random integer number between 1 and $(v-1)$, where v is number of cells. This number defines the crossover point.

2. Swap the groups to the right of the crossover point to generate two offspring.
3. Repair the offspring by eliminating any duplicated machines and introducing missing machines.

Until (selection *poolsize* is achieved).

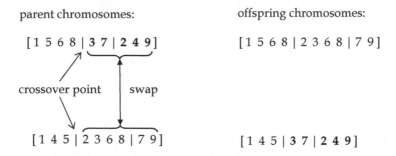

parent chromosomes:

[1 5 6 8 | 3 7 | 2 4 9]

crossover point   swap

[1 4 5 | 2 3 6 8 | 7 9]

offspring chromosomes:

[1 5 6 8 | 2 3 6 8 | 7 9]

[1 4 5 | 3 7 | 2 4 9]

Fig. 3. Crossover operator

In the crossover process, some machines may appear in more than one cell, and some may be missing. Such offspring should be repaired. The repair procedure identifies duplicated machines and eliminates those to the left of the crossover point. Missing machines are inserted on the cell with the least number of nodes. Thus, the group representation scheme enhances the crossover operator by taking advantage of the group structure.

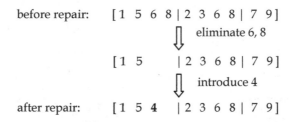

before repair: [1 5 6 8 | 2 3 6 8 | 7 9]

                              eliminate 6, 8

                   [1 5 | 2 3 6 8 | 7 9]

                              introduce 4

after repair: [1 5 4 | 2 3 6 8 | 7 9]

Fig. 4. Chromosome repair procedure

4.5.2 Mutation

The mutation operator is applied to every new chromosome in order to maintain diversity of the population and avoid premature convergence. Two mutation operators are proposed, namely *swap mutation* and *shift mutation*. The swap mutation operates by swapping genes between two randomly chosen groups in a chromosome (see Figure 5.). Its general procedure can be summarized as follows:

1. Randomly select two integer numbers from the set {1,2, ..., v}, where v is the number of cells or groups.
2. Randomly choose a gene from each group

3.   Swap the selected genes, exchanging their values

swap

offspring chromosome	:	[1 5 4	2 3 6 8	7 9]
select group or cell	:	1 and 3
select genes or machines	:	genes 4 and 7
mutated offspring	:	[1 5 7	2 3 6 8	4 9]

Fig. 5. Swap mutation

The *shift mutation operator* works by shifting the frontier between two adjacent groups by one step either to the right or to the left, as shown in Figure 6. Essentially, the number of nodes is increased in one group and simultaneously decreased in the other. The procedure for the mutation operator is summarized thus;

1.   Generate a random integer number between 1 and (v-1). Let this number represent the chosen frontier.
2.   Randomly choose the direction of shift: *right* or *left*.
3.   Shift the frontier in the selected direction, thereby moving one node between adjacent groups.

offspring chromosome	:	[1 5 4	2 3 6 8	7 9]
select frontier, rand (1,2)	:	1 ↳ shift frontier
select direction	:	right
mutated offspring	:	[1 5 4 **2**	3 6 8	7 9]

Fig. 6. Shift mutation operator

4.5.3 Inversion operator

In order to curb premature convergence and control diversity level of the population, an inversion operator is designed. The inversion operator is applied, at a very low probability, on chromosomes selected by the selection strategy prior to crossover operation. Basically, the inversion strategy operates by rearranging the groups in the reverse order, for instance, the order of cells [1, 2, 3] is transformed to a new [3, 2, 1]. This procedure is further illustrated in Figure 7.

chromosome before inversion	:	[1 5 6 8	3 7	2 4 9]
chromosome after inversion	:	[2 4 9	3 7	1 56 8]

Fig. 7. Inversion operator

4.5.4 Diversification

In GGA application it is observed that as the iterations proceed, the solution space (population) converges to a particular solution. However, rapid loss of diversity and premature convergence may occur before an optimal solution is obtained; a problem called genetic drift. To track the diversity of the solution space, Grefenstette (1987) proposed an entropic measure H_i in a population of candidates. For each machine i, the H_i can be defined for GGA in this form;

$$H_i = \sum_{j=1}^{m} \frac{(n_{ij}/p) \cdot \log(n_{ij}/p)}{\log(m)} \tag{8}$$

Where n_{ij} is the number of strings in which machine i is assigned position denoted by j in the current population, p is the solution space size, and m is the number of machines. Divergence H is calculated as;

$$H = \sum_{i=1}^{m} H_i / m \tag{9}$$

As the iterations proceed, the divergence parameter H approaches zero. Thus, the diversity of the solution space can be monitored and controlled by applying the inversion operator till diversity improves to a preset value. In order to prevent loss of good solutions, a fraction (e.g., 0.2) of best performing solutions from the undiversified population is preserved. Performing candidate solutions from the diversified population are compared with those from the diversified population, preferring those that fair better. Thus, the best performing candidates are taken into the next generation.

4.6 The group genetic algorithm implementation

The structure of the proposed GGA for solving the integrated cellular manufacturing system problem was developed incorporating the group operators described in previous sections. A multi-objective approach is adopted in this application based on the two performance measures, ACFI and OFI. The overall GGA structure is now summarised as follows:

Step 1. *Input*: initial data input:

    i. Select the typical initial GGA parameter values (see Table 2)
    ii. Input the manufacturing data, with sequence data

Step 2. *Initial population*: create randomly, two initial populations, called old populations, *oldpop1 and oldpop2*.

Step 3. *Selection and recombination*: Select chromosomes using stochastic sampling without replacement

    i. Evaluate strings by objective function, fitness function and expected count
    ii. create two temporal population, *temppop1* and *temppop2* using the integer parts of expected count and fractional parts as success probabilities

Step 4. *Crossover/recombination*: Apply the group crossover to *temppop1* and *temppop2* to create a two selection pool populations, *spool1* and *spool2*.

    i.   Select two candidates for crossover using remainder selection without replacement, one from *temppop1* and another from *temppop2*.

    ii. Apply crossover operator to the two strings

    iii. If crossover is successful, apply inversion operator, otherwise go to step 5

    iv. Apply repair mechanism if necessary

Step 5.   *Mutation*: apply mutation operators to the two offspring and move them to new population

Step 6.   *Replacement strategy*: Replace old populations with corresponding new populations

    i.   Compare corresponding chromosomes successively in each selection pool and old population

    ii. Take the one that fares better in each comparison

    iii. For the rest of the offspring, selection with probability 0.555

Step 7.   *Diversification*: Diversify population by applying the mutation operator if mutation falls below a predetermined minimum

    i.   Calculate diversity H, of the population

    ii. If the acceptable diversity H_a is such that $H<H_a$ then diversify until diversity is acceptable.

    iii. re-evaluate chromosomes in terms of fitness functions, defined by ACFI and OFI

Step 8.   *New population*: Check the current generation count, *gen* against maximum generation count *maxgen*.

    i.   If *gen* < *maxgen* then go to Step 3, otherwise stop

    ii. Return the best solutions

GGA Parameter	Variable	Value
Number of generations	*maxgen*	Variable
Population size	*popsize*	10-40
Crossover probability	*crossprob*	0.4 – 0.7
Mutation probability	*mutprob*	0.02 – 0.3
Inversion probability	*invprob*	0.04 – 0.2
Chromosome size	*chrom*	Number of machines

Table 2. Typical GGA genetic parameters

Part families are identified based on the number of operations required by a part in a cell. Therefore, a part is assigned to a cell where it requires the maximum number of operations (or machines) for its processing.

5. Results and discussion

The proposed GGA approach was implemented in Java SE 7. An illustration of the GGA execution is first given. A comparative analysis on of the performance of the proposed approach with other algorithms is then presented based on computational analysis on known published data sets.

5.1 GGA computational analysis

This section first provides a numerical illustration obtained when executing the GGA algorithm on well known problem data sets in literature. The set of input data used in this illustration is found in Nair and Narendran (1998). The design and layout problem consists of 25 machines and 40 parts (a 25 x 40 problem). Figure 8 shows an illustration of the intermediate stages arrived at as the algorithm solves design and layout problem. The objective function represents the ACFI and the OFI objective values. The input number of cells used for the simulation run was four. The results of the simulation run show that the ACFI values increased from 20% to 68% after 40 iterations, while the OFI values rose from 21% to 42% after 25 iterations.

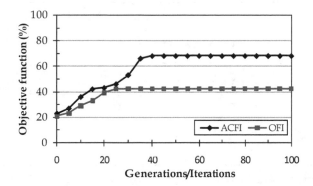

Fig. 8. GGA objective function for Nair & Narendran (1998) 25 x 40 problem

Further numerical experiments were carried out based on an 8 x 20 problem obtained from Nair and Narendran (1998), as shown in Table 3. With a typical set of input data for genetic parameters, the final solution from the GGA simulation run is an improved version of the Nair and Narendran (1998) problem. Table 4 provides the improved solution to the problem. Furthermore, a summary of the final improved version of the solution is provided in Table 5.

	Parts																				
	2	8	9	11	13	14	16	17	19	3	4	6	7	18	20	1	5	10	12	15
	Machines																				
3	2	2	3	2	2	3	2	1	2											
1	1	1	1	3	1	1	1	3	1	2										
4										5	2	2	2	1	1			2		
7			1							3	3	3	3	4	4				2	
8										4	4	4	1	3	5					
2						2				1	1	1	4	2	2					
5							2				5					2	2	3	1	1
6			2												3	1	1	1	3	2

Table 3. Solution from Nair and Narendran (1998) – 8 x 20 problem

The machine cells obtained by the GGA approach are the same as those obtained from the CASE algorithm in Nair and Narendran (1998) and those obtained from the CLASS algorithm in Mahdavi and Mahadevan (2008). Similar to the results from the CLASS algorithm, GGA obtained an improved sequence of machines based on the use of sequence data, showing a remarkable improvement in the layout of machines within cells. In this respect, the GGA approach is effective when compared to well known algorithms in literature. Thus, the algorithm is able to simultaneously address the cell formation and the cell layout problems effectively within a reasonable computation time.

	Parts																			
	2	8	9	11	13	14	16	17	19	3	4	6	7	18	20	15	1	5	10	12
Machines																				
1	1	1	1	3	1	1	1	3	1	2										
3	2	2	3	2	2	3	2	1	2											
2						2				1	1	1	4	2	2					
4										5	2	2	2	1	1				2	
7			1							3	3	3	3	4	4					2
8										4	4	4	1	3	5					
6			2												3	2	1	1	1	3
5						2						5				1	2	2	3	1

Table 4. New solution of Nair and Narendran (1998) using GGA – 8 x 20 problem

Cell	Machines	Parts
C1	1, 3	2, 8, 9, 11, 13, 14, 16, 17, 19
C2	2, 4, 7, 8	3, 4, 6, 7, 18, 20
C3	6, 5	1, 5, 10, 12, 15

Table 5. Final improved solution from Nair and Narendran (1998) problem 8 x 20 problem

Cell No.	CASE Solution				CLASS Solution				GGA Solution			
	n_c	N_{fc}	N_{tc}	CFI%	n_c	N_{fc}	N_{tc}	CFI%	n_c	N_{fc}	N_{tc}	CFI%
1	9	1	9	11.1	9	5	9	55.6	9	5	9	55.6
2	6	7	18	38.9	6	9	18	50	6	9	18	50
3	5	1	5	20.0	5	2	5	40.0	5	2	5	40.0
N_{flow} = 41												
ACFI (%)				21.0				50.0				50.0
OFI (%)				22.0				39.0				39.0

Table 6. Comparative study of GGA, CASE and CLASS algorithms - 8 x 20 problem

In order to demonstrate the utility of the proposed GGA algorithm, a comparative study was done with GGA, CASE and CLASS algorithms. Table 6 provides the results of the comparative analysis. It can be seen from this analysis that though machine groups and part families are the same for the three algorithms, the ACFI and OFI differ with the CASE solution. However, the ACFI and OFI values of GGA are similar to those obtained from CLASS. This shows a remarkable improvement of the solution to the joint cell formation and layout problem.

The next section provides a comparative analysis of the performance of GGA approach and other algorithms found in literature.

5.2 Comparison of GGA with other algorithms

In order to gain more understanding on the effectiveness of the GGA, further comparative experiments were carried out based on data sets reported in literature including Tam (1988), Harhalakis et al. (1990), and Nair & Narendra (1998). Park and Suresh (2003) made a comparative study of known algorithms on sequence data. Algorithms such as fuzzy ART neural network and conventional clustering methods were compared. In addition to these algorithms, other approaches such as CASE designed by Nair and Narendran (1998) and CLASS originated by Mahdavi and Mahadevan (2008) are included in the comparative study. Therefore, the performance of GGA can sufficiently be analyzed based on these known data sets and algorithms. The results obtained in this comparative study are shown in Table 7.

Data set	Size	CLASS			Fuzzy Art			Hierarchical			GGA		
		Cells	ACFI	OFI	Cells	ACFI	OFI	Cells	ACFI	OFI	Cells	ACFI	OFI
1.	12 X 19	2	65%	50%	2	49%	36%	2	48%	45%	2	65%	50%
2.	20 x 20	4	65%	41%	4	42%	34%	4	42%	34%	4	69%	43%
3.	25 X 40	4	52%	34%	7	38%	27%	8	37%	22%	4	68%	42%
4.	08 x 20	3	50%	39%							3	50%	39%

Key: 1. Tam (1988); 2. Harhalakis et al. (1990); 3. Nair & Narendra (1998); 4. Nair & Narendra (1998);

Table 7. A comparison of GGA with other approaches

In all cases, the ACFI and OFI values obtained by GGA are much more preferable than those obtained from other algorithms. From this analysis, it can be seen that the utilization of sequence data in joint cell design and layout is important.

6. Conclusions

Integrated cellular manufacturing system design and layout is an important but hard and complex decision process that involves two main problems; cell formation and machine layout within each cell. Sequence data provides additional information on the dominant flow patterns in cells, which forms the basis for solving the integrated layout problem. However, sequence data has not been fully utilised in manufacturing cell design. The main challenge, therefore, is the extension of the application of sequence data and the development of a robust meta-heuristic algorithm for solving the joint design and layout problem.

In this chapter, a GGA meta-heuristic approach was proposed to solve the integrated layout design problem based on sequence data. The proposed GGA meta-heuristic has unique enhanced features, including a group chromosome scheme, a group crossover operator, a group mutation operator, and a chromosome repair mechanism. The group operators enable the algorithm to reveal the group structure inherent in a data set, producing comparably good quality solutions. While crossover operator enhances exploration of unvisited points in the potential solution space, the mutation exploitation of the best solution in the near-optimal space. Although increasing the number of cells and/or machines may demand more iterations/generations before the algorithm converges to a good solution, the number of parts has no effect on the solution space when grouping machines. Moreover, the parallel mechanism of the approach gives the algorithm robustness and effectiveness over a variety of ill-structured input matrices. Thus, the algorithm is quite preferable in problem situations with a large number of parts.

Comparison with known algorithms in literature was done using known data sets. Apart from well-known performance measures, the average cell flow index was included as a performance parameter, which is a measure of the average magnitude of consecutive forward flows. This measure enabled the GGA approach to evaluate and solve the cell formation and layout design problem in an integrated fashion. The computational results in this study show the utility of the enhanced GGA approach.

Prospects for further research and application of the proposed GGA approach may be interesting. For instance, the group genetic algorithm can be extended to similar clustering problem domains, scheduling problems, as well as network design problems. Further research in these areas is worth exploring.

7. References

Cern, V. (1985). A thermodynamic approach to the travelling salesman problem: An efficient simulation. *Journal of Optimization Theory and Applications*, Vol. 45, 41-51.

Falkenauer E. (1992). The grouping genetic algorithms - widening the scope of the GAs. *Belgian Journal of Operations Research, Statistics and Computer Science*, Vol. 33, 79-102.

Falkenauer, E., 1998, *Genetic Algorithms for Grouping Problems*, New York, Wiley.

Filho, E.V.G., and Tibert, A.J. 2006. A group genetic algorithm for the machine cell formation problem. *International Journal of Production Economics*.

Gen, M. and Cheng, R. (1997). *Genetic Algorithms and Engineering Design*, Wiley Interscience Publication, MA, 1997.

Glover, F. (1989). Tabu Search – Part 1. *ORSA Journal of Computing*, Vol. 1 (2), 190-206.

Glover, F. (1990). Tabu Search – Part 2. *ORSA Journal of Computing*, Vol. 2, No.1, 4-32.

Glover, F., McMillan, C. (1986). The general employee scheduling problem: an integration of MS and AI. *Computers and Operations Research*, Vol. 13, No. 5, 563-573.

Goldberg, D. E. (1989). *Genetic Algorithms: In Search, Optimization & Machine Learning*, Addison-Wesley, Inc., MA, 1989.

Gravel, M., Nsakanda, A.L., Price, W., 1998. Efficient solutions to the cell-formation problem with multiple routings via a double-loop genetic algorithm. *European Journal of Operations Research*, Vol. 109, 286-298.

Grefenstette, J.J. (1987). Incorporating problem specific knowledge into genetic algorithms. In L. Davis, *Genetic and Simulated Annealing*, London: Pitman.

Holland, J. H. (1975). *Adaptation in Natural and Artificial System*, University of Michigan Press, Ann Arbor, MI, 1975.

Hsu, C.M., and Su, C.T. (1998). Multi-objective machine-component grouping in cellular manufacturing: a genetic algorithm approach. *Production Planning & Control*, Vol. 9, No. 2, 155-166.

Jayaswal, S. and Adil, G. K. (2004). Efficient algorithm for cell formation with sequence data, machine replications and alternative process routings. *International Journal of Production Research*, Vol. 42, 2419-2433.

Kaebernick H. and Bazargan-Lari, M. (1996). An integrated Approach to the Design of Cellular Manufacturing. *Annals of the CIRP*, Vol. 45, No. 1, 421-425.

Kirkpatrick, S., Gelatt, C.D., and Vecchi, M.P., (1983). Optimisation by simulated annealing. *Science*, Vol. 220, 621-630.

Kumar, K. R., Kusiak, A. and Vaneli, A. (1986). Grouping parts and components in FMS. *European Journal of Operational Research*, Vol. 24387-24400.

Mahdavi, I. and Mahadevan, B. (2008). CLASS: An algorithm for cellular manufacturing system and layout design using sequence data. *Robotics and Computer-Integrated Manufacturing*, Vol. 24, 488-497.

Man, K. F., Tang, K. S. and Kwong, S. (1999). *Genetic Algorithms: Concepts and Design*, Springer, London, 1999.

Nair, G. J. and Narendran, T. T. (1998). CASE: a clustering algorithm for cell formation with sequence data. *International Journal of Production Research*, Vol. 36, 157-179.

Onwubolu, G.C. and Mutingi, M. (2001). A genetic algorithm approach to cellular manufacturing systems. *Computers & Industrial Engineering*, Vol. 39, 125-144.

Sarker B. R., and Xu, Y. (1998). Operation sequences-based cell formation methods: a critical survey. *Production Planning & Control*, Vol. 9, 771--783.

Selim, H. M., Askin, R. G., and Vakharia, A. J. (1998). Cell formation in group technology: Evaluation and directions for future research. *Computers & Industrial Engineering*, Vol. 34, No. 1, 3--20.

Singh N. (1993). Design of cellular manufacturing systems: an invited review. *European Journal of Operational Research*, Vol. 69, 284--291.

Suresh NC, Slomp J, KAparth S., 1999. Sequence-dependent clustering of parts and machines: a Fuzzy ART neural network approach. *International Journal of Production Research*, Vol. 37, 2793-816

Venugopal, V., Narendran, T.T., 1992. A genetic algorithm approach to the machine-component grouping problem with multiple objectives. *Computers and Industrial Engineering*, Vol. 22, No. 4, 469-480.

Won, Y. and Lee, K. C. (2001). Group technology cell formation considering operation sequences and production volumes. *International Journal of Production Research*, Vol. 39, 2755-2768.

9

Platform for Intelligent Manufacturing Systems with Elements of Knowledge Discovery

Tomasz Mączka and Tomasz Żabiński
Rzeszów University of Technology
Poland

1. Introduction

Numerous and significant challenges are currently being faced by manufacturing companies. Product customization demands are constantly growing, customers are expecting shorter delivery times, lower prices, smaller production batches and higher quality. These factors result in significant increase in complexity of production processes and the necessity for continuous optimization. In order to fulfil market demands, managing the production processes require effective support from computer systems and continuous monitoring of manufacturing resources, e.g. machines and employees. In order to provide reliable and accurate data for factory management personnel the computer systems should be integrated with production resources located on the factory floor.

Currently, most production systems are characterized by centralized solutions in organizational and software fields. These systems are no longer appropriate, as they are adapted to high volume, low variety and low flexibility production processes. In order to fulfil current demands, enterprises should reduce batch sizes, delivery times, and product life-cycles and increase product variety. In traditional manufacturing systems this would create an unacceptable decrease in efficiency due to high replacements costs, for example. (Christo & Cardeira, 2007)

Modern computer systems devoted to manufacturing must be scalable, reconfigurable, expandable and open in the structure. The systems should enable an on-line monitoring, control and maximization of the total use of manufacturing resources as well as support human interactions with the system, especially on the factory floor. Due to vast amounts of data collected by the systems, they should automatically process data about the manufacturing processes, human operators, equipment and material requirements as well as discover valuable knowledge for the factory's management personnel. The new generation of manufacturing systems which utilizes artificial intelligence techniques for data analyses is referred to as Intelligent Manufacturing Systems (IMS). IMS industrial implementation requires computer and factory automation systems characterized by a distributed structure, direct communication with manufacturing resources and the application of sophisticated embedded devices on a factory floor. (Oztemel, 2010)

Many concepts in the field of organizational structures for manufacturing have been proposed in recent years to make IMS a reality. It seems that the most promising concepts are: holonic (HMS or Holonic Manufacturing System), fractal, and bionic manufacturing. Further references can be found in (Christo & Cardeira, 2007). In general, it could be stated that a promising organizational structure is a conglomerate of distributed and autonomous units which operate as a set of cooperating entities. It would be impossible to successfully implement the new organizational concepts in the manufacturing industry without suitable distributed control and monitoring hardware and software.

In publications (Leitão, 2008), (Colombo et al., 2006) an agent-based software is designated as technology for industrial IMS realization, regardless of the chosen organizational paradigm. The integration of HMS and multi-agent software technology is currently presented as the most promising foundation for industrial implementations of IMS. The HMS paradigm is based on concepts originally developed by Arthur Koestler in 1969 with reference to social organizations and living organisms. The term holon describes a basic unit of organization and has two important characteristics: autonomy and co-operation. In a manufacturing system, a holon can represent a physical or logical activity, e.g. a machine, robot, order, machine section, flexible manufacturing system and even a human operator. The holon possesses the knowledge about itself and about the environment and has an ability to cooperate with other holons. From this viewpoint, a manufacturing system is a holarchy, which is defined as a system of holons organized in a dynamical hierarchical structure. Manufacturing system goals are achieved by cooperation between holons. Due to conceptual similarities of HMS and agent-based software, it seems clear that their combination should create a promising platform for an industrial IMS implementation.

On the other hand, humans still play an important role in manufacturing systems and in spite of predictions from the seventies, which suggested that human operators would no longer be needed in fully automated production, they even play a more important role nowadays than they did in the past (Oborski, 2004). Proper cooperation between humans and machines or humans and manufacturing control systems can significantly improve overall production effectiveness, so human system interface design is still an active research area. Human System Interfaces (HSI) are responsible for efficient cooperation between operators and computer systems. (Gong, 2009) It seems clear that convenient and reliable human system interaction is a key factor for successful industrial IMS implementation.

In this chapter, results of the project devoted to the development of a hardware and software platform for IMS are to be described. The platform based on Programmable Automation Controllers (PAC) has already been successfully tested by being included in everyday production processes in four small and medium Polish metal component production companies. The platform was employed in order to monitor production resources in real-time and to conduct communication between computer systems, machines and operators. On the basis of the tests results, it has been experimentally proven that the main development-related barrier for real deployment of IMS (Leitão, 2008), i.e. absence of industrial controllers with appropriate capabilities, is out-of-date. Within 28 months of the system operation it has been proven that modern PAC are capable of running data processing, communication and graphical user interface modules directly on the PAC controller in parallel with PLC programs. The hardware and software system which has been created constitutes of a platform for future complete implementation of the IMS.

The project has been created by the Department of Computer and Control Engineering, Rzeszów University of Technology, in cooperation with Bernacki Industrial Services company and the students from the Automation and Robotics scientific circle called ROBO (ROBO, 2011). The project has been made under the auspices of Green Forge Innovation Cluster.

The long-term goal of this project is the development and industrial implementation of full IMS concept for small and medium production companies. The goal determines two main assumptions for the selection of hardware and software elements. The first assumption is a possibility to include in the system newer machines with advanced control equipment as well as older ones without controllers. The second one is a reasonable cost of the system. In the paper (Żabiński et al., 2009) the results of the first stage of the project were presented. During the first stage, the hardware was selected and the prototype and functionality limited testbed for one machine was installed in a screw manufacturing factory. The system was tested in a real production process. The first stage of the project was to provide valuable benefits for the factory management board in the field of monitoring availability, performance and quality of operation of the machines and operators. Additionally, control tasks in a PLC layer of the system were done using ST (Structured Text) programming language for the machines which had not been previously equipped with a controller. The success of the first stage resulted in the project continuation and the findings of the current stage are presented in this paper and in (Mączka & Czech, 2010), (Żabiński & Mączka, 2011), (Mączka & Żabiński 2011), (Mączka et al., 2010).

The main goal of the current project stage is to develop and test in a real production environment a system structure with PACs integrated with touchable panels for each machine. Some improvements of HSI for the factory floor are being made, according to suggestions acquired during the first stage, and new functionalities for different organizational units, i.e. maintenance, tool and material departments. The important part of current and future works is to employ artificial intelligence and data mining technology to give factory management personnel reliable and long-term knowledge about the production processes.

2. Systems testbed structures

Up to now, four system industrial testbeds have been constructed. The first one was installed in a screw manufacturing company which is a member of the Green Forge Innovation Cluster. The Green Forge Innovation Cluster is an association of metal production companies and scientific institutions from southern Poland, which aims at innovative solutions development for metal components production. The second one was installed at the WSK PZL-Rzeszów company, in the department which produces major rotating parts for the aviation industry. The first testbed consists of two machine sections formed by cold forging press machines. The first section includes machines without PLC controllers but the second one consists of 12 modern machines equipped with PLCs and advanced cold forge process monitoring devices. The second testbed includes one production line with four CNC vertical turning lathes and two CNC machining centres. The machines in the testbeds have been operated by experienced operators who interact with the system using various peripheral devices, such as barcode readers, electronic calipers and industrial touch panels. (Żabiński & Mączka, 2011) There is also a third testbed installed in a different department at "WSK PZL-Rzeszów", which monitors one machine.

During the project, a mobile testbed with GSM communication was also constructed. The purpose of a mobile testbed is to allow production companies to test the system without bearing the costs of communication infrastructure installation. Thanks to such a testbed, companies can better define system functionalities better which are very important for them, taking into account the production profile and organizational structure. Currently, this testbed has been installed in a metal component manufacturing company, where one machine has been monitored. (Mączka & Żabiński, 2011)

In the hardware and software part of the system, three main layers can be distinguished: a factory floor hardware and software, a data server layer and WWW (World Wide Web) client stations.

2.1 Factory floor layer

Due to the scalable, reconfigurable, expandable and open structure of the platform, the industrial implementations differ in functionality as well as in hardware and software elements installed on a factory floor.

2.1.1 Testbed with PACs

In the first implementation type, Programmable Automation Controllers (PACs), also known as embedded PCs, are used on a factory floor. PACs are equipped with operating systems and meet the demands of modern manufacturing systems as they combine features of traditional Programmable Logic Controllers (PLCs) and personal computers (PCs). The main feature of PACs is the ability to use the same device for various tasks simultaneously, e.g. data acquiring, processing and collection; process and motion control; communication with databases or information systems; Graphical User Interface (GUI) implementation, etc. There are two kinds of Windows system available for the controllers, i.e. Windows CE and Windows XP Embedded. Windows CE is equipped with .Net Compact Framework, Windows XP Embedded is equipped with .Net Framework. There are benefits of using the XP Embedded platform, e.g. homogeneity of the software platform for controllers and PC stations as well as availability of network and virus protection software. Due to the financial reasons, Ethernet network for communication and controllers with Windows CE were chosen for the two testbeds.

In the system, PACs acquire data form machines using distributed EtherCAT (Granados, 2006) communication devices equipped with digital or analog inputs. Each PAC is equipped with Windows CE 6.0 operating system, real-time PLC subsystem, UPS (Uninterruptible Power Supply), Ethernet as well as RS-232/485 interfaces for communication and DVI (Digital Visual Interface)/USB (Universal Serial Bus) interfaces for touchable monitors connection. One controller with peripherals, i.e. an industrial 15″ touch panel, RFID (Radio-frequency Identification) cards reader and barcode reader, is installed in each machine section or production line. The hardware system structure for a factory floor is shown in Fig. 1.

The software for PACs consist of two layers. The first layer is PLC software written in ST (Structured Text) language, which is mainly responsible for reading and writing physical inputs and outputs. The second layer constitutes the application written in C# language for .NET Compact Framework (CF), which runs under Windows CE. (Microsoft Developer Network, 2011)

The second layer consists of the following modules:

- the module for communication between the PLC program and other system parts,
- the module for communication with database using web services technology,
- the operator's GUI.

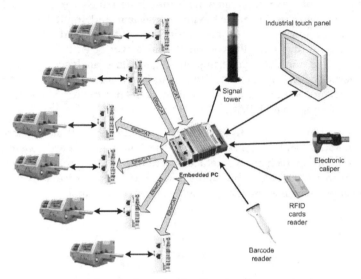

Fig. 1. Hardware structure of the platform for the first implementation type

PLC control programs run in the PLC layer on the same device simultaneously with GUI, data processing and database communication modules which run in Windows CE layer (Fig. 2). The ADS (Automation Device Specification) protocol enables C# programs to read and write data directly from and to PLC programs via names of PLC variables. It significantly simplifies the communication between PLC and C# applications.

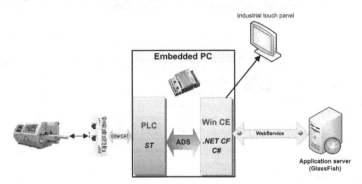

Fig. 2. PAC software structure

The PLC and Windows software for embedded PC controller was designed and implemented in order to control up to 6 machines. It gives flexibility in the system structure, e.g. for more demanding PLC or CNC control tasks there is a configuration of one controller for one possible machine. For simple machines or machines already equipped with controllers, it is possible to create a configuration with one embedded PC controller and up to six machines. This configuration can be used, for instance, to incorporate machine sections into the system.

Currently, the new implementation type supported by EU funds, is under construction. It concerns installation of PACs integrated with touchable panels for each machine (Fig. 3). The panel allows operators to interact with the system and input basic data regarding corresponding machines, e.g. reason for production stoppage and references for orders, materials and tools. In this case, the computational power of each PAC is devoted to one machine and is used for data gathering, intelligent data analysis, intelligent condition monitoring, alarms detection, etc. The PACs models were selected in order to provide sufficient performance for the future multi-agent system structure version, with separate machine agents running on each PAC.

In this implementation type, apart from separate PAC for each machine, personal computers with HSI for operators can be flexibly connected to the system. PCs are treated as additional system "access points", which allow interaction with the system on the same level as PACs. Additionally, it can e.g. provide technical documentation for a particular production process in a more convenient way than on the PAC with a small touchpanel.

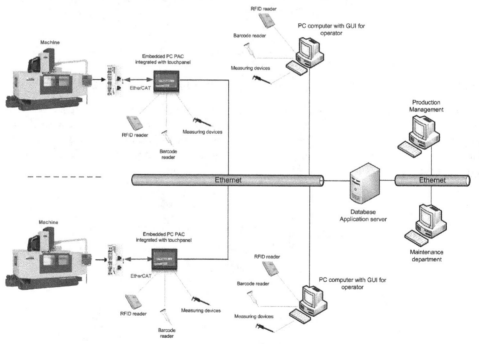

Fig. 3. Hardware structure of the platform with PAC for single machine

2.1.2 Testbed with industrial PC

System structure in the "WSK PZL-Rzeszów" testbed is going to be modified, in order to obtain more flexible architecture. The new structure should also simplify the system implementation in companies with various production and data resources. The new software structure should enable its easy adaptation to the needs of other production sectors. During the development of the testbed, additional diagnostics and process monitoring equipment will be included in the system, e.g. quality measurement devices, current and force sensors, etc. In concern to the WSK testbed, it is planned to include an additional 64 machines in the system. Due to customer demands, one industrial PC computer will be installed for each production line. The industrial PC's task is monitoring states of the machines included in particular production line on base of digital (machine work mode, machine engine state) and analog (spindle load) signals from machines' control systems. Windows XP Embedded with real-time and soft PLC TwinCAT subsystem is the operating environment for industrial PC; connection with input/output modules is performed via EtherCAT bus using star network topology. Chosen topology minimizes communication problems if some part of the EtherCAT network infrastructure will fail, e.g. in case of network cable break.

In this type of implementation, machine operators interact with the system using PCs workstations, placed near monitored machines. The basic scope of data, which can be viewed and input carried out by operators, is similar like in PAC implementation. Detailed description will be given in section 3.1.

Additionally, the system is going to be integrated with an SAP business software and will be used for delivering electronic versions of technical and quality control documents directly to operators' workstations. The system should support production management as well as support the maintenance department by the usage of advanced and intelligent machine condition monitoring software tools. The new testbed structure is shown in Fig. 4. (Żabiński & Mączka 2011)

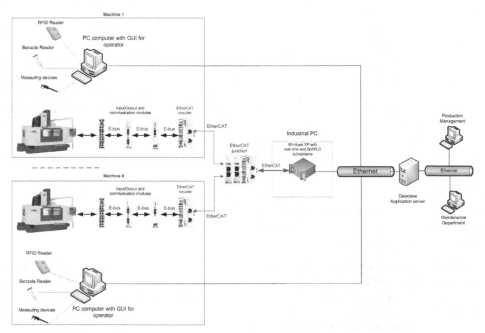

Fig. 4. Hardware structure with industrial PC

2.2 Data server and end clients layer

The data and application server layer is common for all the structures. It includes the PostgreSQL database server and the GlashFish application server. GlassFish is an open-source application server compatible with Java Platform Enterprise Edition (J2EE). (GlassFish Community, 2011) PostgreSQL is open source object-relational database system which conforms to the ANSI-SQL (Structured Query Language) 2008 standard. (PostgreSQL, 2011) The application server hosts communication and data processing modules with web services written in Java. The WWW client layer utilizes websites written in JSF (Java Server Faces), JSP (Java Server Pages), AJAX (Asynchronous Javascript and XML) and also works under GlassFish server control. Communication between PACs or PCs and the database and between the presentation layer and the database is performed using web services or Enterprise Java Beans (EJB) technology.

3. Human system interfaces

Human System Interface, which was developed for the IMS project, consists of two main layers, i.e. a factory floor layer and a WWW layer. The first layer is a GUI application which runs on an embedded PC installed on the factory floor. In this layer the communication between an operator and the system is done via a 15″ touchable monitor. The second layer is a web page accessed through a web browser from the factory intranet or the Internet. The Polish language is used in HSI, as the system has been installed in Polish factories. Due to this reason, the GUI language presented in this section is Polish.

3.1 Factory floor layer

The HSI for factory floor layer has two main operation modes, i.e. locked and unlocked. In the locked mode an operator can only observe information presented on the screen. In the unlocked mode an operator can interact with the system. An operator can change the HSI mode using his RFID card. Thanks to the RFID operator's badges a security policy was implemented on the factory floor. In the locked mode visual information of machines operation mode, production plan and plan realization and the necessity of an operator interaction with the system is presented. The necessity of an operator interaction with the system is indicated by the blinking of a panel associated with the machine which needs intervention. The locked mode screen for a machine section with six machines is presented in Fig. 5.

In the unlocked HSI mode an operator can perform various tasks connected with the system, e.g. login, logout, taking up shift, order selection or confirmation, stop reason and quality control data input etc. The unlocked mode screen for a machine section with six machines is presented in Fig. 6.

The HSI consists of two main sections, i.e. the system section (it allows login, logout, etc.) and the machines section (it allows data input for particular machines). Small rectangles with letters O, SR, QC, S, associated with each machine panel (a large rectangle with machine name, e.g. Tłocznia T-19), indicates the action which should be performed for the particular machine, i.e. O – order selection or confirmation, SR – stop reason input, QC – quality control data input and S – service. When the machine panel is blinking, an operator can quickly determine the operation which should be performed, the color of the letters O, QC, SR or S becomes red for the active action. An example of an order input screen is shown in Fig. 7. Operators can input order code using on-screen keyboard or via barcode reader. (Żabiński & Mączka, 2011)

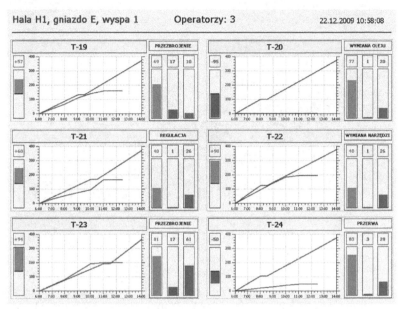

Fig. 5. The factory floor HSI locked mode screen for six machines

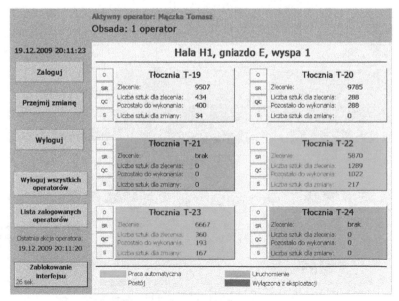

Fig. 6. The factory floor HSI unlocked mode screen for six machines

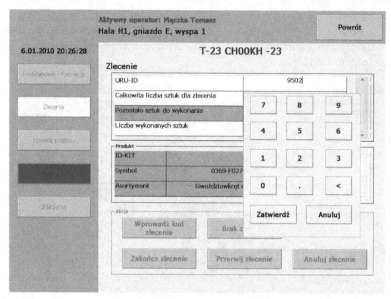

Fig. 7. Order inputting screen for particular machine

3.2 WWW layer

The WWW layer includes two main modules, i.e. an on-line view and statistics. The on-line view enables on-line monitoring of machines operation mode, e.g. production, stoppage, lack of operator and also other information like: operator identifier, order identifier, shift production quantity, daily machine operation structure or detailed history of events. The on-line view for a production hall is presented in Fig. 8.

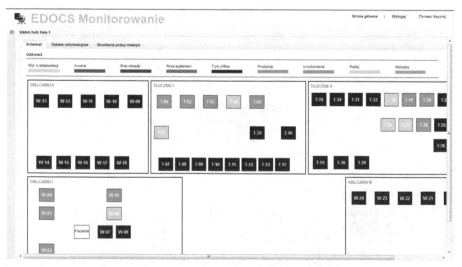

Fig. 8. Production state on-line view for a production hall

The statistics module enables computing some statistical factors concerning: machines work time, production quantity, failures, orders, operators work time etc. It enables users to configure statistics parameters, i.e. analysis time interval, statistics elements, type of presented data (for instance daily or weekly analysis type) and chart type (e.g. bar graph, line graph). An example of a statistics screen for machine production quantity with its configuration options is shown in Fig. 9. (Żabiński & Mączka, 2011)

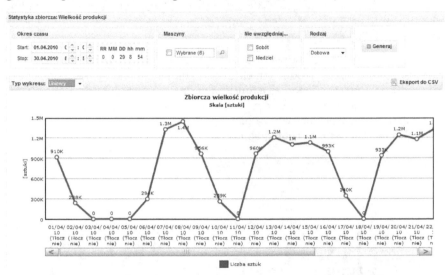

Fig. 9. Machines production quantity statistic with its configuration options

4. System operation results

Currently, four system testbeds are installed in real production environments in small and medium factories and have been used in daily production processes. All of the testbeds currently utilize the first structure described in section 2.1.1, with embedded PCs on the factory floors. The first testbed has been installed in a screw manufacturing factory since 16 May, 2009. Eighteen machines for cold forging are currently monitored. The second testbed was installed at the WSK PZL-Rzeszów company, in the department which produces major rotating parts for the aviation industry and it has been in operation since 21 September, 2010 with six CNC machines included in the system. The new system implementation with the second type structure for this testbed, described in section 2.1.2, is under construction. There is also one testbed where a single machine is monitored using a mobile (all-in-one) system testbed with GSM communication. The next separated testbed with one machine was installed in April 2011 in a different department at "WSK PZL-Rzeszów"of the aviation parts producing company.

Currently, the system is responsible for collecting data concerning machine operation and operators' work. The PLC layer is responsible for detecting and registering events which occurred in the machines, for instance the oil pump and the main motor start/stop, failure and emergency signals, the machine operational mode (manual or automatic), signals from diagnostics modules (process monitoring devices), etc. The PLC program is also responsible for registering the quantity produced. Information about events, including timestamps, machine and operator identifiers and other additional parameters, is stored in the database. Two mechanisms are used to store data in the database, such as: an asynchronous event driven method and a synchronous one with a 10 second time period for diagnostics purposes. The system is also responsible for detecting and storing information on breakdowns, setup and adjustments, minor stoppages, reduced speed etc. Every production stoppage must be assigned with an appropriate reason. Tool failure, for example, are automatically detected, while others have to be manually chosen by operators via the HSI.

On the server side, there are software modules used for calculating different KPIs (Key Performance Indicators), e.g. production efficiency, equipment and operators efficiency etc. The real production quantity report as a function of day, calculated for time interval from 1-03-2011 to 31-03-2011 for 6 machines, is presented in Fig. 10. During this period, the planned production time for the machines was 24 hours per day (3 shifts). As shown in Fig. 10, there were some fluctuations of production quantity.

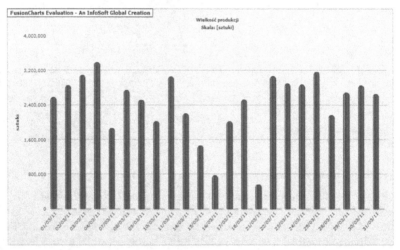

Fig. 10. Production quantity report – number of produced items as a function of days

A machine operation time structure is analyzed and can be shown as a horizontal graph (Fig. 11). At the moment there is a possibility of analyzing data from three points of view: a general view, a view with stop reasons and a detailed view. The general view divides machine operational time into three categories, i.e. the operator's absence, the automatic production and the stoppage. In the analysis for the view with stop reasons, each stoppage time period is associated with the appropriate stop reason. In the detailed view, periods of the manual machine operation are distinguished in each stoppage time. Different colors are designated for appropriate time intervals (Fig. 11), e.g. the stoppage – light brown, the automatic production – green, the manual operation – dark green, the start-up time – blue, the electrical breakdown – red, etc.

During the long term test, when the system was included in the regular daily production, it was proven that the selected hardware and software platform is suitable for industrial implementation of the IMS. The software modules (HSI, communication, web services, data acquisition) for Windows CE have been running successfully on the embedded PC controller in parallel to PLC programs. (Mączka & Czech, 2010)

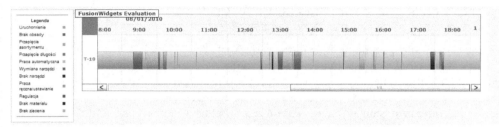

Fig. 11. Machine operation time structure with stop reasons

Within 704 days of system operation in the screws production testbed, the whole number of events registered in the testbed was 3,883,374. The system covers: production monitoring, quality control, material and tools management and also fundamental support for the maintenance department. In the company which produces aviation parts, 57,230 events were registered within 59 days of system operation.

5. Association rules application for knowledge discovery

Taking into account the limited number of machines included in the currently working system testbeds, it can be stated that the number of data collected is considerable. A long-term data analysis to make reasoning and generalized conclusions about production processes would be a demanding task for human analysts. Therefore, there is a need to employ artificial intelligence and data mining technology to give factory management personnel reliable knowledge of the production processes. In this section, the results of the initial tests in the areas of applying data mining and artificial intelligence techniques to discover knowledge about the production processes, are to be described.

It is expected that continuously discovered knowledge will support everyday production process management and control, thus providing the answers to numerous questions, e.g. what are the bottlenecks in the production systems etc. Moreover, it is envisaged that the system will be able to automatically identify relationships in the production systems, discover possibilities for more effective usage of production resources and use Statistical Process Control (SPC) with artificial intelligence support for early detection of possible problems in production systems.

Initial work in this area concerns the creation of tools for automatic rules (patterns) generation, which will describe relations between values in the database. The discovered patterns could be used for detecting operators' improper actions, which could have an influence on machine operation, e.g. increasing downtime duration and number of breakdowns. The rules are referred to as association rules.

5.1 Introduction to the experiment

The goal of association rules is to detect relationships between specific values of categorical variables in large data sets.

The formal definition of the problem is as follows: Let $D = \{t_1; t_2; ...; t_m\}$ be a set of m transactions, called data set or database. Let $I = \{i_1; i_2; ...; i_n\}$ be a set of possible n binary attributes for transaction, called items. Single transaction T is a set of items such that $T \subseteq I$.

Assume that X is a set of some items from I, so $X \subseteq I$. A transaction T contains X if the transaction contains all items from X, so $X \subseteq T$.

An association rule is an implication of the form X => Y, where $X \subset I$, $Y \subset I$, and $X \cap Y=\varnothing$. The rule X => Y holds in the transaction set D with confidence c if c% of transactions in D that contain X also contain Y. The rule X => Y has support s in the transaction set D if s% of transactions in D contain $X \cup Y$. Given a set of transactions D, the problem of mining association rules is to generate all association rules that have support and confidence greater than the user-specified minimum support (minsup) and minimum confidence (minconf) respectively. (Agrawal & Srikant, 1994).

The experiment of finding association rules in production data has been performed. The first step of the experiment was to choose the subset of data to analyze, i.e. time period and attributes, and to extract raw data from screw manufacturing company database to CSV (Comma Separated Values) text format. CSV format was chosen because of the possibility of loading data directly into data mining software, i.e. Statistica or Weka. 127232 events registered on 12 machines from 4.01.2011 to 16.07.2011, concerning machine operational state, were extracted using SQL query and pgAdmin database management tool. The structure of events is shown in Table 1, the table contains only a subset of registered events.

plc_time	event_type_id	machine_id
2011-01-04 07:59:25.784	2000	41
2011-01-04 08:26:27.565	2001	41
2011-01-04 08:28:49.845	2000	41
2011-01-04 08:29:42.705	2001	41
2011-06-14 21:49:38.032	2000	44
2011-06-14 23:29:17.732	2001	44
2011-07-11 06:59:49.812	2000	68
2011-07-11 07:35:15.332	2001	68
2011-07-11 07:41:44.872	2000	68
2011-07-11 08:21:07.812	2001	68

Table 1. Structure of raw events extracted from production database

Attributes of a single event are:

- plc_time– time of event registration in the PLC layer
- event_type_id– type of registered event, 2000 is production start, 2001 is production stop
- machine_id– identifier of a machine for which the event was registered

After consultation with the company production management personnel, an assumption has been made that length of times of continuous machine state, i.e. length of production and length of stoppage will be important factor to analyze. It seems to be clear that if continuous production time of a particular machine lasts longer than the others, this machine works more efficiently, without the need for operator action. It is worthwhile noticing that lower number of stoppages should positively affect machine lifetime and save energy.

5.2 Data preparation

The format for raw data presentation in Table 1 is not useful for discovering associations concerning machines continuous interval length, as raw events do not directly reflect particular machine state during a particular period of time. Because of this, data needs to be pre-processed to the list of production and stoppage intervals for each machine. Each record should contain interval start date, interval end date and interval length. Pre-processing task was done using Python script, which analyzes events list and produces stoppage intervals, if the current analyzed event is 2001 (production stop) and next analyzed event is 2000 (production start). In the opposite situation, production interval is inserted to the result list. Data structure after the pre-processing phase is shown in Table 2.

start	End	type	length_min	machine_id
2011-01-04 07:59:25.784	2011-01-04 08:26:27.565	W	27.03	41
2011-01-04 08:26:27.565	2011-01-04 08:28:49.845	S	2.37	41
2011-01-04 08:28:49.845	2011-01-04 08:29:42.705	W	0.88	41
...				
2011-06-14 21:49:38.032	2011-06-14 23:29:17.732	W	99.66	44
...				
2011-07-11 06:59:49.812	2011-07-11 07:35:15.332	W	35.43	68
2011-07-11 07:35:15.332	2011-07-11 07:41:44.872	S	6.49	68
2011-07-11 07:41:44.872	2011-07-11 08:21:07.812	W	39.38	68

Table 2. Data structure after the pre-processing phase

Attributes of single interval:

- start – timestamp of interval begin,
- end – timestamp of interval end,
- type – interval type, W is work, S is stoppage,
- len_min – interval time length in minutes,
- machine_id – identifier of machine associated with interval.

Start and end timestamps has only informational role and they are omitted in the process of finding associated rules. However, data presented in Table 2 are not ready for application of associated rules finding algorithms. It results from fact, that known association rules discovering algorithms deals with data, whose attributes have discrete or categorical values. In above case, attributes *type* and *machine_id* are categorical, but *len_min* has continuous values. The solution of this problem is mapping attributes with continuous values to categorical attributes, referred in (Agrawal & Srikant, 1996) as partitioning quantitative attributes.

This transformation was, like the previous, performed by Python script. Number of categories and length of each category's interval were chosen based on minimum and maximum values of *len_min*, in order to obtain regular distribution of data in generated categories. 46 categories were generated, starting from [0-0.1m] (interval length greater than 0 to 0.1 minutes, or 10 seconds) to [700-+Inf m] (interval length greater of equal 700 minutes). Example values of processed items are contained in table 3. Letter 'M' was added before machines identifiers to indicate that this is categorical, not numerical attribute.

start	End	type	length_category	machine_id
2011-01-04 07:59:25.784	2011-01-04 08:26:27.565	W	[20.1-30.1m]	M41
2011-01-04 08:26:27.565	2011-01-04 08:28:49.845	S	[2.1-3.1m]	M41
2011-01-04 08:28:49.845	2011-01-04 08:29:42.705	W	[0.8-0.9m]	M41
...				
2011-06-14 21:49:38.032	2011-06-14 23:29:17.732	W	[90.1-100.1m]	M44
...				
2011-07-11 06:59:49.812	2011-07-11 07:35:15.332	W	[30.1-40.1m]	M68
2011-07-11 07:35:15.332	2011-07-11 07:41:44.872	S	[6.1-7.1m]	M68
2011-07-11 07:41:44.872	2011-07-11 08:21:07.812	W	[30.1-40.1m]	M68

Table 3. Data after partitioning quantitive attribute *len_min* to categorical attribute *length_category*

5.3 Finding association rules using Weka

For finding associations in previously prepared data, Weka (Waikato Environment for Knowledge Analysis) software was used. This is a popular suite of a machine learning software written in Java that was developed at the University of Waikato, which is available under the GNU General Public License. Weka is a collection of machine learning algorithms for data mining tasks. Weka contains tools for not only for association rules, but also for classification, regression, clustering and visualization. Its architecture is well-suited for developing new machine learning schemes. (Hall et al., 2009)

In the first step, Weka knowledge explorer was run and preprocessed mining dataset was loaded from CSV file. Weka displays list of attributes in dataset and its basic information like categories, number of items in each category etc. (Fig. 12).

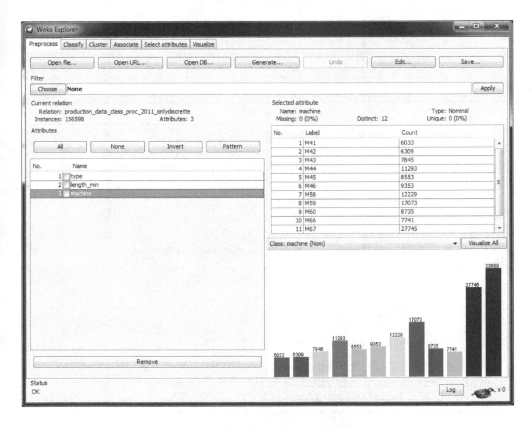

Fig. 12. Weka initial screen after data loading

The next step was running module for association rules generation, choosing algorithm for mining rules and configuring algorithm's parameters. Weka provides few rules discovering algorithms, including Apriori, Predictive Apriori, Generalized Sequential Patterns. In the described experiment, most popular Apriori algorithm was selected with parameters:

support 0.1 (10%), confidence 0.3 (30%). The small value of support was choosen experimentally to not omit rules for particular machines, and purpose of 30% confidence was to ignore irrelevant rules, which applies only to relative small part of data set. Configuration parameters are shown in Fig. 13.

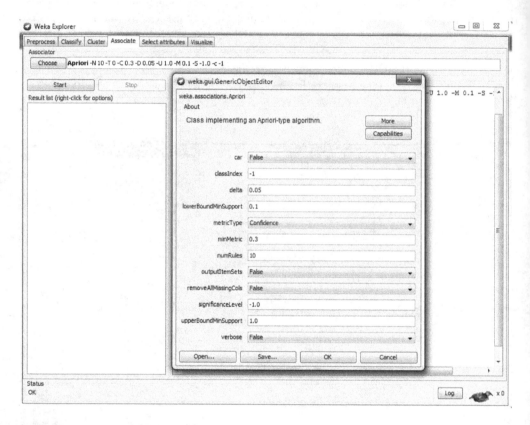

Fig. 13. Apriori algorithm configuration

The algorithm found 10 rules listed below:

1. machine=M67 [27745] ==> length_min=0-0.1m [20099] conf:(0.72)
2. machine=M68 [33689] ==> length_min=0-0.1m [23825] conf:(0.71)
3. length_min=0-0.1m [64730] ==> type=W [37527] conf:(0.58)
4. machine=M68 [33689] ==> type=W [16845] conf:(0.5)
5. machine=M68 [33689] ==> type=S [16844] conf:(0.5)
6. type=W [78305] ==> length_min=0-0.1m [37527] conf:(0.48)
7. length_min=0-0.1m [64730] ==> type=S [27203] conf:(0.42)
8. length_min=0-0.1m [64730] ==> machine=M68 [23825] conf:(0.37)
9. type=S [78293] ==> length_min=0-0.1m [27203] conf:(0.35)
10. length_min=0-0.1m [64730] ==> machine=M67 [20099] conf:(0.31)

Number of records which contain attributes values listed in predecessor and consequent are present in bracket squares. Some potentially interesting rules are underlined, their interpretation may be as follows.

Rules 1 and 2 indicates that for machine M67 and M68, production or stoppage interval lasts usually relatively short - from above 0 to 10 seconds. It suggests that listed machines may have some troubles with stable work, which can lead to low efficiency. Potential reasons of this situation are mechanical problems, improper material, improper operator's actions etc. This situation is also covered by rules 8 and 10, where predecessor and consequent are reversed, these rules are more general.

On the other hand, 48% of work intervals and 35% of stoppage intervals lasts up to 10 seconds, so short time intervals can be, in some level, screw production profile and pushing process specific. However, such potential problem appears to be interesting for deeper analyze by experts from production manager personnel.

5.4 Finding association rules with Statistica

The same experiment like in 5.3 was performed in Statistica 10 Data Miner. This commercial software includes modules for neural network, clusterization, classification trees etc. (StatSoft, 2011) For the association rules mining, module called basket analysis is available. It includes only one rules finding algorithm, the same like used in previous experiment, i.e. Apriori.

Rules found by Statistica, shown in Fig. 14, are the same like in previous experiment, where non-commercial Weka software was used.

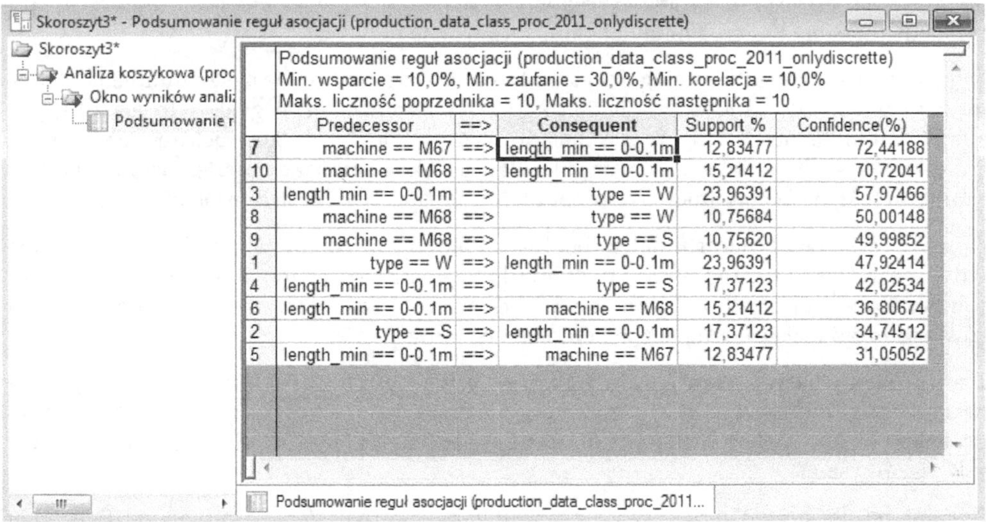

	Predecessor	==>	Consequent	Support %	Confidence(%)
7	machine == M67	==>	length_min == 0-0.1m	12,83477	72,44188
10	machine == M68	==>	length_min == 0-0.1m	15,21412	70,72041
3	length_min == 0-0.1m	==>	type == W	23,96391	57,97466
8	machine == M68	==>	type == W	10,75684	50,00148
9	machine == M68	==>	type == S	10,75620	49,99852
1	type == W	==>	length_min == 0-0.1m	23,96391	47,92414
4	length_min == 0-0.1m	==>	type == S	17,37123	42,02534
6	length_min == 0-0.1m	==>	machine == M68	15,21412	36,80674
2	type == S	==>	length_min == 0-0.1m	17,37123	34,74512
5	length_min == 0-0.1m	==>	machine == M67	12,83477	31,05052

Fig. 14. Association rules found by Statistica's basket analysis module

6. Conclusion

So far the project devoted to preparing platform for industrial implementation of Intelligent Manufacturing System, the hardware and software has been selected and tested in a real production process using the preliminary testbeds. Currently, the platform is mainly used for data collection concerning the production processes, machine operation and operators' work. It also provides HSI for machines' operators and system end-users, i.e. factory management board.

The hardware and software layer, which has already been installed in the factory, has created the need and the basis for the employment of advanced data mining and artificial intelligence techniques and multi-agent software structure in the system. Such techniques could be used for detecting operators' improper actions which could have an influence on machines operation, e.g. increasing downtime duration and number of breakdowns. The analysis of the rules could probably give an answer to the following questions, e.g. what factors and in what manner influence production processes, under what circumstances problems occur, how operators react to diagnostic messages, etc. So far, experiment of finding association rules from the data gathered by production monitoring system was performed for the fixed range of events concerning 12 machines work. Some potentially interesting rules, which can help the factories management personnel to detect and eliminate bottlenecks in the production processes, were discovered. Process of finding association rules is not currently automated, but thanks to open source data mining software Weka, it can be easily integrated with the existing system infrastructure. Remaining work to be done in this area includes extending events attributes list, by adding for each item in data set product type identifier and machine's operators identifier. Experiments with different types of rules discovering algorithms, i.e. Apriori Predictive and Generalized Sequence Patterns, will be also performed. Eventually, the system should automatically generate rules, which help to detect the possibility of the problem occurrence on the basis of historical data. Additionally it should suggest the best solution to the problem, therefore the probability of stoppages will be reduced.

New system structures are being developed in order to simplify the system implementation in companies with various production and data resources. The goal is to achieve easy adaptation to the needs of many production sectors. Some of these structures can decrease the cost of system deployment, as popular devices like personal computers can be used on condition that the factory floor is appropriately adaptable, e.g. if there is no oil mist.

However, it should be clearly stated that for industrial implementation of IMS, the structure with separate PAC installed for each machine seems to be the most promising. PAC computational power allows running separate machine agents for each machine, so one of new organizational structure, e.g. holonic, can be implemented.

Tests of new system versions, especially those with multi-agent solutions, are going to be performed with the usage of the laboratory Flexible Manufacturing System (FMS) testbed, presented in Fig. 15. The FMS testbed consists of an integrated CNC milling machine, robot and vision system.

Current results of the project give promising perspectives for an advanced Intelligent Manufacturing System development and implementation with PACs as hardware platform in a real factory for next project stages.

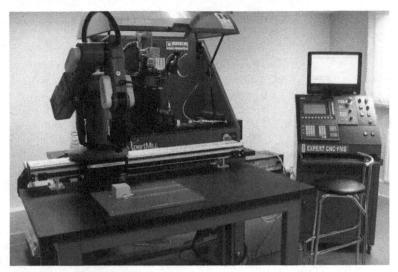

Fig. 15. Laboratory FMS testbed

7. Acknowledgment

FMS testbed was bought as a part of the project No POPW.01.03.00-18-012/09 from the Structural Funds, The Development of Eastern Poland Operational Programme co-financed by the European Union, the European Regional Development Fund.

8. References

Agrawal R., Srikant R. (1994). Fast Algorithms for Mining Association Rules in Large Databases. *VLDB'94, Proceedings of 20th International Conference on Very Large Data Bases*, pp. 487-499, ISBN 1-55860-153-8, Morgan Kaufmann, Santiago de Chile, Chile, 1994.

Agrawal R., Srikant R. (1996). Mining quantitative association rules in large relational tables. *Proceedings of the 1996 ACM SIGMOD international conference on Management of data*, pp. 1 - 12, ISBN 0-89791-794-4, ACM New York, NY, USA, 1996.

Christo C., Cardeira C. (2007). Trends in Intelligent Manufacturing, *Proceedings of IEEE International Symposium on Industrial Electronics*, pp. 3209-3214, ISBN 978-1-4244-0754-5, Vigo, Spain, June 4-7, 2007.

Colombo A., Schoop R., Neubert R. (2006). *An Agent-Based Intelligent Control Platform for Industrial Holonic Manufacturing Systems*, IEEE Trans. Ind. Elect., vol. 53, no. 1, pp. 322-337, ISSN: 0278-0046, Seligenstadt, Germany, February 2006.

GlassFish Community (2011). *GlassFish Server Open Source Edition*, 23.11.2011, Available from: http://glassfish.java.net/.

Gong C. (2009). *Human-Machine Interface: Design Principles of Visual Information*. Human-Machine Interface Design. In: Proc IEEE Conference on Intelligent Human-Machine Systems and Cybernetics, pp 262–265, San Antonio Texas, USA, 2009.

Granados F. (2006). *Analysis: Industrial Ethernet - Driving the growth,* Computing & Control Engineering Journal, vol.17, no.6, pp.14-15, Dec.-Jan. 2006.

Hall M., Frank E., Holmes G., Pfahringer B., Reutemann P., Witten I. (2009). *The WEKA Data Mining Software: An Update,* pp.10-18, SIGKDD Explorations, Volume 11, Issue 1, ACM New York, NY, USA, 2009.

Leitão P. (2008). Agent-based distributed manufacturing control: A state-of-the-art survey. *Engineering Applications of Artificial Intelligence,* No. 22 (7), pp. 979-991, ISBN 0952-1976, ELSEVIER, 2008.

Microsoft Developer Network (2011). *C# Language Specification,* 23.11.2011, Available from http://msdn.microsoft.com/en-us/library/aa645596(v=vs.71).aspx.

Mączka T., Czech T. (2010). Manufacturing Control and Monitoring System – Concept and Implementation. *Proceedings of IEEE International Symposium on Industrial Electronics,* t.1, p.3900-3905, Bari, Italy, July 4-7 2010.

Mączka T., Żabiński T. (2011). System for remote machines and operators monitoring - selected elements (in Polish). *Pomiary Automatyka Robotyka,* p. 62-65, No. 3/2011.

Mączka T., Czech T., Żabiński T (2010). Innovative production control and monitoring system as element of factory of future (in Polish). *Pomiary Automatyka Robotyka,* p.22-25, No. 2/2010.

Oborski P. (2004). Man-machine interactions in advanced manufacturing systems. *The International Journal of Advanced Manufacturing Technology,* Vol. 23. No. 3-4, pp 227-232, ISSN: 1433-3015, Springer-Link, 2004.

Oztemel E. (2010). Intelligent manufacturing systems. *L. Benyoucef, B. Grabot, (Ed.) Artificial Intelligence Techniques for Networked Manufacturing Enterprises Management,* pp. 1-41, ISBN 978-1-84996-118-9, Springer-Verlag, London, 2010.

PostgreSQL (2011). *About,* 23.11.2011, Available from: http://www.postgresql.org/about/.

ROBO (2011). *Student Automation and Robotics scientific circle ROBO (in Polish),* 23.11.2011, Available from http://www.robo.kia.prz.edu.pl/.

StatSoft (2011). *STATISTICA Product Overview,* 23.11.2011, Available from http://www.statsoft.com/products/.

Żabiński T., Mączka T. (2011). Implementation of Human-System Interface for Manufacturing Organizations. *Human-Computer Systems Interaction. Backgrounds and Applications 2,* Advances in Soft Computing, Springer-Verlag Co., 2011.

Żabiński T., Mączka T., Jędrzejec B. (2009). Control and Monitoring System for Intelligent Manufacturing – Hardware and Communication Software Structure. *Proceedings of Computer Methods and Systems,* p. 135-140, Kraków, Poland, November 26-27 2009.

10

Hybrid Manufacturing System Design and Development

Jacquelyn K. S. Nagel[1] and Frank W. Liou[2]
[1]James Madison University
[2]Missouri University of Science and Technology
USA

1. Introduction

Reliable and economical fabrication of metallic parts with complicated geometries is of considerable interest for the aerospace, medical, automotive, tooling, and consumer products industries. In an effort to shorten the time-to-market, decrease the manufacturing process chain, and cut production costs of products produced by these industries, research has focused on the integration of multiple unit manufacturing processes into one machine. The end goal is to reduce production space, time, and manpower requirements. Integrated systems are increasingly being recognized as a means to meet these goals. Many factors are accelerating the push toward integrated systems. These include the need for reduced equipment and process cost, shorter processing times, reduced inspection time, and reduced handling. On the other hand, integrated systems require a higher level of synthesis than does a single process. Therefore, development of integrated processes will generally be more complex than that of individual unit manufacturing processes, but it could provide simplified, lower-cost manufacturing.

Integrated systems in this research area have the ability to produce parts directly from a CAD representation, fabricate complex internal geometries, and form novel material combinations not otherwise possible with traditional subtractive processes. Laser metal deposition (LMD) is an important class of additive manufacturing processes as it provides the necessary functionality and flexibility to produce a wide range of metallic parts (Hopkinson et al., 2006; Liou & Kinsella 2009; Venuvinod & Ma, 2004). Current commercial systems that rely on LMD to produce tooling inserts, prototype parts, and end products are limited by a standard range of material options, building space, and a required post-processing phase to obtain the desired surface finish and tolerance. To address the needs of industry and expand the applications of a metal deposition process, a hybrid manufacturing system that combines LMD with the subtractive process of machining was developed achieving a fully integrated manufacturing system.

Our research into hybrid manufacturing system design and development has lead to the integration of additive and subtractive processes within a single machine footprint such that both processes are leveraged during fabrication. The laser aided manufacturing process (LAMP) system provides a rapid prototyping and rapid manufacturing infrastructure for research and education. The LAMP system creates fully dense, metallic parts and provides all the advantages of the commercial LMD systems. Capabilities beyond complex geometries and good surface finish include: (1) functional gradient material metallic parts where different materials are added from one layer to the next or even from one section to another, (2) seamlessly embedded sensors, (3) part repair to reduce scrap and extend product service life, and (4) thin-walled parts due to the extremely low processing forces (Hopkinson et al., 2006; Liou et al., 2007; Ren et al., 2008). This hybrid system is a very competitive and economical approach to fabricating metallic structures. Hybrid manufacturing systems facilitate a sustainable and intelligent production model and offer flexibility of infrastructure to adapt with emergent technology, customization, and changing market needs (Westkämper, 2007). Consequently, the design strategies, system architecture, and knowledge required to construct hybrid manufacturing systems are vaguely described or are not mentioned at all in literature.

The goal of this chapter is to summarize the key research findings related to the design, development, and integration of a hybrid manufacturing process that utilizes LMD to produce fully dense, finished metallic parts. Automation, integration, and control strategies along with the associated issues and solutions are presented as design guidelines to provide future designers with the insight needed to successfully construct a hybrid system. Following an engineering design perspective, the functional and process knowledge of the hybrid system design is explored before physical components are involved. Key results are the system architecture, qualitative modeling, and quantitative modeling and simulation of a hybrid manufacturing process.

In summary, this chapter provides an interdisciplinary approach to the design and development of a hybrid manufacturing system to produce metal parts that are not only functional, but also processed to the final desired surface-finished and tolerance. The approach and strategies utilized in this research coalesce to facilitate the design of a sustainable and intelligent production system that offers infrastructure flexibility adaptable with emergent technologies and customizable to changing market needs. Furthermore, the approach to hybrid system design and development can assist in general with integrated manufacturing systems. Applying the strategies to design a new system or retrofit older equipment can lead to increased productivity and system capability.

2. Related work

Any process that results in a solid physical part produced directly from a 3D CAD model can be labeled a rapid prototyping process (Kalpakjian & Schmid, 2003; Venuvinod & Ma, 2004). Equally, a process that converts raw materials, layer-by-layer into a product is a rapid prototyping process, but is typically referred to as additive manufacturing or layered manufacturing. Subtractive manufacturing is the process of incrementally removing raw material until the desired dimensions are met. Where additive processes start from the ground up, subtractive processes start from the top down. The combination of manufacturing processes from different processing categories establishes a hybrid manufacturing system. Herein, a hybrid manufacturing system refers to a manufacturing system that is comprised of an additive and subtractive manufacturing process.

Both additive and subtractive manufacturing cover a wide range of fabrication processes. For example, additive manufacturing can involve powder-based (e.g., selective laser sintering), liquid-based (e.g., stereolithography) or solid-based (e.g., fused deposition modeling) processes, each using a wide range of materials (Gebhardt, 2003; Kai & Fai, 1997; Venuvinod & Ma, 2004). While traditional subtractive manufacturing is typically reserved for metals, advanced or non-conventional subtractive processes have emerged to handle a greater variety of materials which include electric discharge machining, water jet cutting, electrochemical machining and laser cutting (Kalpakjian & Schmid, 2003). The physical integration of additive and subtractive manufacturing processes, such as laser metal deposition and machining, is the key to leveraging the advantages of each process. The vast domains of additive and subtractive manufacturing have provoked many to test boundaries and try a new concept, in an attempt to discover the next best system that will play a key role in advancing manufacturing technologies. Academic and industry researchers alike have been developing novel, hybrid manufacturing systems, however, the design and integration strategies were not published. On the other hand, a few approaches taken to develop reliable hybrid systems that deliver consistent results, with the majority based on consolidation processes, have published a modest guide to their system design. In following paragraphs, a number of hybrid manufacturing systems are reviewed to give an idea of what has been successful.

Beam-directed technologies, such as laser cladding, are very easy to integrate with other processes. Most have been integrated with computer numerically controlled (CNC) milling machines by simply mounting the cladding head to the z-axis of the milling machine. Kerschbaumer and Ernst retrofitted a Röders RFM 600 DS 5-axis milling machine with an Nd:YAG laser cladding head and powder feeding unit, which are all controlled by extended CNC-control (Kerschbaumer & Ernst, 2004). Similarly, a Direct Laser Deposition (DLD) process utilizing an Nd:YAG laser, coaxial powder nozzle and digitizing system as described by (Nowotny et al., 2003) was integrated into a 3-axis Fadal milling machine. Laser-Based Additive Manufacturing (LBAM) as researched at Southern Methodist University, is a technique that combines an Nd:YAG laser and powder feeder with a custom built motion system that is outfitted with an infrared imaging system (Hu et al., 2002). This process yields high precision metallic parts with consistent process quality. These four systems perform all deposition steps first, and then machine the part to the desired finish, consistent with conventional additive fabrication.

Two powder-based manufacturing processes that exhibit excellent material usage and in most cases produced components do not require finishing are Direct Metal Laser Sintering (DMLS) and Laser Consolidation (LC). Using layered manufacturing technology, a DMLS system such as the EOS EOSINT M270 Xtended system, can achieve an acceptable component finish using a fine 20 micron thick metal powder material evenly spread over the build area in 20 micron thick layers (3axis, 2010). Laser Consolidation developed by NRC Canada is a net-shape process that may not require tooling or secondary processing (except interfaces) (Xue, 2006, 2008). Parts produced using these processes exhibit net-shape dimensional accuracy and surface finish as well as excellent part strength and material properties.

Non-conventional additive processes demonstrate advanced features, alternate additive and subtractive steps, filling shell casts, etc. A hybrid RP process proposed by (Hur et al., 2002) combines a 6-axis machining center with any type of additive process that is machinable, a sheet reverse module, and an advanced process planning software package. What differentiates this process is how the software decomposes the CAD model into machining and deposition feature segments, which maximize the CNC milling machine advantages, and significantly reduces build time while increasing shape accuracy. Laser welding, another hybrid approach, involves a wire feeder, CO_2 laser, 5-axis milling center, and a custom PC-NC based control unit that has been used to produce molds for injection molding (Choi et al., 2001). Hybrid-Layered Manufacturing (HLM) as researched by (Akula & Karunakaran, 2006) integrates a TransPulse Synergic MIG/MAG welding process with a conventional milling machine to produce near-net shape tools and dies. This is direct rapid tooling. Welding and face milling operations are alternated to achieve desired layer height and to produce very accurate, dense metal parts. A comparable process was developed at Fraunhofer IPT named Controlled Metal Build-up (CMB), in which, after each deposited layer the surface is milled smooth (Kloche, 2002). However, CMB utilizes a laser integrated into a conventional milling machine.

Song and Park have developed a hybrid deposition process, named 3D welding and milling because a wire-based gas metal arc welding (GMAW) apparatus has been integrated with a CNC machining center (Song & Park, 2006). This process uses gas metal arc welding to deposit faster and more economically. Uniquely, 3D welding and milling can deposit two materials simultaneously with two welding guns or fill deposited shells quickly by pouring molten metal into them. The mold Shape Deposition Manufacturing (SDM) system at Stanford also uses multiple materials to deposit a finished part, however, for a different purpose (Cooper, 1999). A substrate is placed in the CNC mill and sturdy material such as UV-curable resin or wax is deposited to form the walls of a mold, which then is filled with an easily dissolvable material. The top of the mold is deposited over the dissolvable material to finish the mold; once the mold has cooled down the dissolvable material is removed, and replaced with the desired part material. Finally, the sturdy mold is removed to reveal the final part, which can be machined if necessary. Contrary to the typical design sequence (Jeng & Lin, 2001) constructed their own motion and control system for a Selective Laser Cladding (SLC) system and integrated the milling head, which evens out the deposition surface after every two layers. Clearly, each system has its advantages and contributes differently to the RM industry.

Although using a CNC milling machine for a motion system is the most common approach to constructing a hybrid system, a robot arm can easily be substituted. This is the case with SDM created at Stanford University (Fessler et al., 1999). The robot arm was fitted with an Nd:YAG laser cladding head which can be positioned accurately, allowing for selective depositing of the material and greatly reducing machining time. Integration of a handling robot can reduce positioning errors and time between operations if the additive and subtractive processes are not physically integrated.

Most of the aforementioned systems have been built with versatility in mind and could be set-up to utilize multiple materials or adapted to perform another operation. However, an innovative hybrid system that has very specific operations and capabilities is the variable lamination manufacturing (VLM-ST) and multi-functional hotwire cutting (MHC) system (Yang et al., 2005). The VLM-ST system specializes in large sized objects, up to 3 ft. x 5 ft., by converting polystyrene foam blocks into 3D objects utilizing the turntable of the 4-axis MHC system during cutting; if the object is bigger still, multiple pieces are cut and put together.

The design strategy behind several of the reviewed hybrid systems was not emphasized and documented. Thus, key pieces of information for the design and development of hybrid systems are missing which prevents researchers and designers from easily designing and constructing a hybrid system of their own. The information contained within this chapter aims to provide a comprehensive overview of the design, development, and integration of a hybrid manufacturing system such that others can use as a guideline for creating a hybrid system that meets their unique needs.

3. Research approach

As previously mentioned, the design strategies, system architecture, and knowledge required to construct a hybrid manufacturing system is vaguely described if mentioned at all in the literature. Consequently, our research approach is mainly empirical. Although our approach relies heavily on observation and experimental data, it has allowed us to identify opportunities for applying theory through modeling and simulation.

A major challenge to hybrid manufacturing system design is accurately controlling the physical dimension and material properties of the fabricated part. Therefore, understanding the interaction of all process parameters is key. Layout of the preliminary system architecture provides a basis for qualitative modeling. Independent and dependent process parameters are identified through qualitative modeling, which defines the parameters that require a quantitative understanding for accurate control of the process output. Qualitative models of the hybrid manufacturing process are developed and analyzed to understand both process and functional integration within the hybrid system. This allowed lost, competing or redundant system functionality to be identified and used to inform design decisions. Modeling how the material and information flows through the hybrid system facilitates the development of the automation, integration, and control strategies.

Quantitative modeling and simulation of our hybrid manufacturing system concentrates on process control and process planning. Process control modeling is used to predict the layer thickness via an empirical model based on the direct 3D layer deposition, the particle concentration of the powder flow, the nozzle geometry, the carrier gas settings, and the powder-laser interaction effects on the melt pool. Process planning models are used to automate part orientation, building direction, and the tool path. These models assist with resolving the challenges of the laser deposition process including building overhang structures, producing precision surfaces, and making parts with complex structures.

Revisiting the preliminary system architecture design with the knowledge gained from qualitative and quantitative modeling has resulted in a system architecture that enables accurate and efficient fabrication of 3D structures. Decomposition of the system architecture allows for direct mapping of customer needs and requirements to the overall system architecture.

4. Hybrid manufacturing system

The laser aided manufacturing process (LAMP) lab at Missouri University of Science and Technology (formerly University of Missouri-Rolla) houses a 5-axis hybrid manufacturing system, which was established by Dr. Liou and other faculty in the late 1990s. This system entails additive-subtractive integration, as shown in Fig. 1, to build a rapid prototyping/manufacturing infrastructure for research and education at Missouri S&T. Integration of this kind was planned specifically to gain sturdy thin wall structures, good surface finish, and complex internal features, which are not possible by a LMD or machining system alone. Overall, the system design provides greater build capability, better accuracy, and better surface finish of structures with minimal post-processing while supporting automated control. Applications of the system include repairing damaged parts (Liou et al., 2007), creating functionally gradient materials, fabrication of overhang parts without support structures, and embedding sensors, and cooling channels into specialty parts.

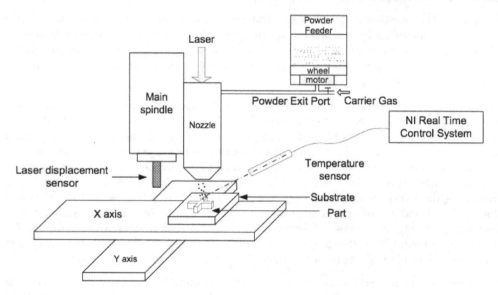

Fig. 1. Five-axis Hybrid Manufacturing Process (Adapted from Tang et al., 2007)

The LAMP hybrid system is comprised of five subsystems or integration elements: process planning, control system, motion system, manufacturing process, and a finishing system. Equipment associated with subsystem is described in the following paragraphs and summarized in Table 1.

The LAMP process planning system is a in-house layered manufacturing or slicing software that imports STL models from a commercial CAD package to generate a description that specifies melt pool length (mm), melt pool peak temperature, clad height (mm) and sequences of operations. The objective of the process planning software is to integrate the five-axis motion and deposition-machining hybrid processes. The results consist of the subpart information and the build/machining sequence (Ren et al., 2010; Ruan et al., 2005). To generate an accurate machine tool path a part skeleton, which calculates distance and offset edges or boundaries, is created of the CAD model. Distance, gradient, and tracing functions were modified to allow more complicated and unconnected known environments for successful implementation with the LAMP hybrid manufacturing system. Basic planning steps involve determining the base face, extracting the skeleton, decomposing a part into subparts, determining build sequence and direction for subparts, checking the feasibility of the build sequence and direction for the machining process, and optimization of the deposition and machining.

Hybrid Manufacturing Subsystems	LAMP Hybrid System Equipment
Process Planning	Commercial and in-house CAD software
Motion	Fadal 3016L 5-axis VMC
Manufacturing Process	Nuvonyx 1kW diode laser, Bay State Surface Technologies 1200 powder feeder
Control	NI RT PXI chassis & LabVIEW, Mikron temperature sensor, Omron laser displacement sensor, Fastcom machine vision system
Finishing	Fadal 3016L 5-axis VMC

Table 1. LAMP Hybrid System Equipment

True 3D additive manufacturing processes can be achieved with a 5-axis machining center without additional support structures (Ruan et al., 2005), as opposed to 2.5D that is afforded by a 3-axis machine. Therefore the motion subsystem for the LAMP hybrid manufacturing system is a 5-axis Fadal 3016L VMC, which also constitutes the finishing subsystem. Servo motors control the motion along the axes as compared with crank wheels and shafts in conventional machine tools. The Fadal VMC is controlled via G and M codes either entered at the control panel or remotely fed through an RS-232 connection.

The main manufacturing process of the hybrid system is laser metal deposition, the additive manufacturing process. Metal powder is melted using a 1kW diode laser while the motion system traverses in response to the tool path generated by the process planning software, thereby creating molten tracks in a layer-by-layer fashion on a metal substrate. Layers are deposited with a minimum thickness of 10μm. The melt pool temperature is between 1000°C and 1800°C, depending on the material (e.g. H13 tool steel, Titanium alloy), but is less than 2000°C. A commercial powder delivery system, designed for plasma-spraying processes carries the steel or titanium powder to the substrate via argon. The cladding head is mounted to the z-axis of the Fadal VMC to fully utilize the motion system and provide the opportunity to machine the fabricated part at any point in the deposition process by applying a translation algorithm. The beam focusing optics, beam splitter for out-coupling the process radiation from the laser beam path, water cooling connections, powder feeder connections, and various sensors (optional) are located within the cladding head. Built in to the cladding head are pathways for metal powder to travel through to the laser beam path in a concentric form, therefore, releasing metal powder in a uniform volume and rate. Quartz glass is used to focus the laser beam and water carried from the chiller to the cladding head by small plastic hoses reduces the wear on the focusing optics. Overall, the LMD subsystem includes equipment for lasing, cooling, and powder material delivery.

Control of the hybrid manufacturing subsystems require a versatile industrial controller and a range of sensors to acquire feedback. The National Instruments Real Time Control System (NI RT System) provides analog and digital I/O ports and channels, DAC, RS-232, and ADC for controlling all the subsystems of the hybrid system. The control system contains a PXI-

8170 Processor, 8211 Ethernet card, 8422 RS-232 card, 6527 Digital I/O card, 6711 Analog Output card, 6602 Timing I/O card, 6040E Multi-function card, and an SCXI Controller with 1304 card. PCI eXtensions for Instrumentation (PXI) is a PC-based platform for measurement and automation systems. PXI combines PCI electrical-bus features with the modular, Eurocard packaging of Compact PCI, and then adds specialized synchronization buses and key software features. Signal Conditioning Extension for Instrumentation (SCXI) is a front-end signal conditioning and switching system for various measurement devices, including plug-in data acquisition devices. Our control system offers modularity, expandability, and high bandwidth in a single, unified platform.

System feedback is acquired through temperature and laser displacement sensors. An Omron Z4M-W100 laser displacement sensor is used to digitally determine the cladding head height above the substrate. There are danger zones and safe zones that the nozzle can be with respect to the substrate. Output of the displacement sensor is -4 to +4 VDC which is converted into a minimum and maximum distance value, respectively. The temperature sensor is a Mikron MI-GA5-LO non- contact, fiber-optic, infrared temperature sensor. It was installed onto the Z-axis of the VMC with a custom, adjustable fixture. The set-up for data acquisition of the melt pool temperature, while deposition takes place is at an angle of 42°, 180 mm from the melt pool and sampling every 2 ms. There is also a machine vision system, a Fastcom iMVS-155 CMOS image sensor, to watch the melt pool in real-time. It has also been used to monitor melt pool geometry and assist with our empirical approach to fine tune process parameters.

5. Hybrid manufacturing system design and development

The critical success factors of an integrated system are quality, adaptability, productivity and flexibility (Garelle & Stark, 1988). Inclusion of additive fabrication technology in a traditional subtractive manufacturing system inherently addresses these four factors. Nevertheless, considering the four success factors during the initial design phase will ensure that the resultant manufacturing system will meet short and long term expectations, be reliable, and mitigate system obsolescence. In order for hybrid manufacturing systems to become a widespread option they must also be an economical solution. Dorf and Kusiakpoint out that the three flows within a manufacturing system i.e. material, information, and cost, which "should work effectively in close cooperation for efficient and economical manufacturing" (Dorf & Kusiak, 1994). This section reviews the qualitative and quantitative modeling efforts of material and information as well as the system architecture design that incorporates the knowledge gained through modeling. Cost modeling for the hybrid system has only been temporal, however, a cost benefit analysis as proposed in (Nagel & Liou, 2010) could be performed to quantify the savings.

5.1 System architecture

Initially, the LAMP system design was integrated only through the physical combination of the laser metal deposition process (additive manufacturing) and the machining center (subtractive manufacturing). Also, each subsystem housed a separate controller, including the LMD and VMC, which required manual control of the hybrid system. Reconfiguring the LAMP hybrid system to utilize a central control system, increased communication between the subsystems and eliminated the need for multiple people. Moreover, the process can be controlled and monitored from a remote location, increasing the safety of the manufacturing process. The hybrid manufacturing system architecture follows the modular, integration element structure as defined in (Nagel & Liou, 2010). Figure 2 shows the direct mapping of customer needs and requirements to the overall system architecture as well as the dependency relationships. Build geometry, surface finish, and material properties are the needs relating directly to the finished product. Efficient operation and flexibility are the system requirements to be competitive and relate directly to the system itself.

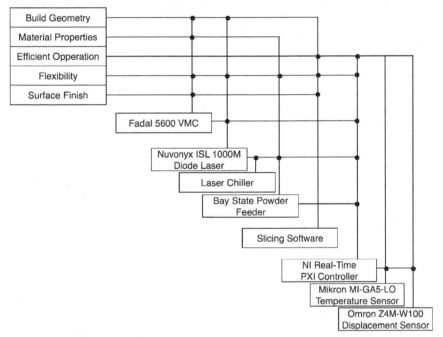

Fig. 2. LAMP Hybrid System Architecture

5.2 Qualitative modeling

Qualitative modeling efforts are focused on understanding process parameters and the flow of the process. Modeling the process parameter interactions uncovers the independent and dependent process parameters where as modeling the manufacturing process identifies opportunities for optimization. The following subsections summarize how qualitative modeling has been used to gain knowledge of the relationships among process parameters and resources utilized in each step of the hybrid manufacturing process.

5.2.1 Independent process parameters

The major independent process parameters for the hybrid manufacturing system include the following: laser beam power, process speed, powder feed rate, incident laser beam diameter, and laser beam path width (path overlap) as shown in Fig. 3 (Liou et al., 2001). Other parameters such as cladding head to surface distance (standoff distance), carrier gas flow rate, absorptivity, and depth of focus with respect to the substrate also play important roles.

Dependent Parameters | **Independent Parameters**

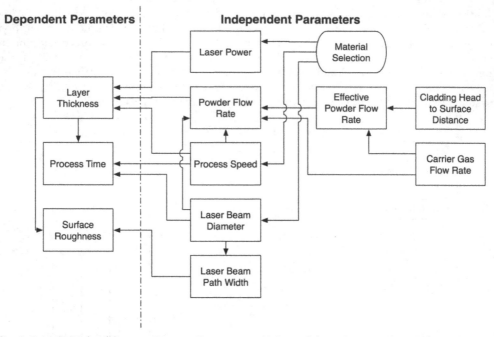

Fig. 3. LAMP Hybrid System Process Parameters (Adapted from Liou et al., 2001)

The layer thickness process parameter is directly related to the power density of the laser beam and is a function of incident beam power and beam diameter. Generally, for a constant beam diameter, the layer thickness increases with increasing beam power provided corresponding powder feed rate. It was also observed that the deposition rate increased with increasing laser power (Weerasinghe & Steen, 1983).

Powder mass flow rate is another important process parameter which directly affects layer thickness. However, effective powder flow rate, which includes powder efficiency during the LMD process, turned out to be a more important parameter (Lin & Steen, 1998; Mazumder et al., 1999). Also the factor that most significantly affected the percent powder utilization was laser power. The cladding head nozzle is set up to give a concentric supply of powder to the melt pool, and due to the nature of the set-up, the powder flow is hour glass-shaped. The powder flow initially is unfocused as it passes through the cladding head, but the nozzle guides the powder concentrically towards its center, and essentially "focuses" the beam of powder. The smallest diameter focus of the powder "beam" is dependent upon the design of the cladding head nozzle. Also, if the laser beam diameter becomes too small as compared to the powder beam diameter, e.g., 100μm, much of the supplied powder will not reach the melt pool. Thus, there will be unacceptably low powder utilization.

Process speed has a big impact on the process output. In general, decreasing process speed increases the layer thickness. There is a threshold to reduce process speed, however, as too much specific energy (as defined in Section 5.2.2) will cause tempering or secondary hardening of previous layers (Mazumder et al., 1997). Process speed should be well chosen since it has strong influence on microstructure.

The laser beam diameter parameter is one of the most important variables because it determines the power density. It can be difficult to accurately measure high power laser beams. This is partly due to the shape of the effective beam diameter (e.g., Gaussian, Top hat) and partly due to the definition of what is to be measured. Single isotherm contouring techniques such as charring paper and drilling acrylic or metal plates are well known but suffer from the fact that the particular isotherm they plot is both power and exposure time dependent. Multiple isotherm contouring techniques overcome these difficulties but are tedious to interpret.

Beam path width or beam width overlap has a strong influence on surface roughness. As the deposition pass overlap increases, the valley between passes is raised due to the overlap therefore reducing the surface roughness. Powder that has adhered to the surface, but has not melted will be processed in successive passes. In order to obtain the best surface quality, the percent pass overlap should be increased as much as possible. Conversely, to decrease the surface roughness, the deposition layers should be kept as thin as possible.

5.2.2 Dependent process parameters

The major dependent process parameters of the hybrid manufacturing system are: layer thickness, surface roughness, and process time (Fig. 3). Other dependent parameters such as hardness, microstructure, and mechanical properties should also be considered, but in this chapter we will focus only on the parameters related with physical dimension.

There is a large range of layer thicknesses as well as deposition rates that can be achieved using LMD. However, part quality consideration puts a limit on optimal deposition speeds. Both the layer thickness and the volume deposition rates are affected predominately by the specific energy and powder mass flow rate. Here, specific energy (SE) is defined as: $SE = p/(Dv)$, where p is the laser beam power, D is the laser beam diameter and v is the process speed. Also it has been well known that actual laser power absorbed in the melt pool is not the same as the nominal laser power measured from a laser power monitor due to reflectivity and other plasma related factors depending on the materials (Duley, 1983). The use of adjusted specific energy is thus preferable. Considering the factors, there is a positive linear relationship between the layer thickness and adjusted specific energy for a range of powder mass flow rates (Liou et al., 2001).

Surface roughness was found to be highly dependent on the direction of measurements with respect to the deposited metal (Liou et al., 2001; Mazumder et al., 1999). In checking the surface roughness, at least four directions should be tested from each sample; the length and width direction on the top surface, and the horizontal and vertical directions on thin walls. Since the largest roughness on each sample is of primary interest, measurements should be only taken perpendicular to the deposition direction on the top surface and in the vertical direction on the walls, based on our experiments.

The overall deposition processing time is mainly dependent upon the layer thickness per slice, process speed, and laser beam diameter. The processing conditions need to be optimized prior to optimizing the processing time, since the processing time is directly influenced by the processing conditions. If the laser beam diameter is increased, the specific energy and power density will be decreased under the same process condition, that means, a lower deposition rate unless the laser power and powder mass flow rate are increased correspondingly. Similarly, when the process speed is increased the independent process parameters should be optimized accordingly.

5.2.3 Process modeling

Process modeling used to model the hybrid manufacturing system aims to optimize the sequence with which the material flows through the system (Shunk, 1992; Wang, 1997). Following the process modeling approach by (Nagel et al., 2009), process events and tasks within each event were identified. Part A of Fig. 4 shows the manually controlled hybrid manufacturing process. Decomposition of the system process aided with identification of integration points to reduce the number of steps and events within the process resulting in significant time savings. Part B of Fig. 4 shows the optimized process.

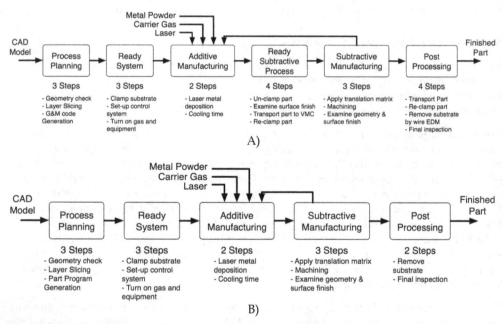

Fig. 4. LAMP Hybrid System Process Models, A) Before integration and optimization, B) After integration and optimization

Once the process is clearly laid out, the motion system and control system can be accurately defined. Reconfiguring the LAMP hybrid system elements to utilize a central control system increased communication between the subsystems and eliminated the need for multiple people. Moreover, the process can be controlled and monitored from a remote location, increasing the safety of the manufacturing process. Supplementary improvements were made to the process planning software, laser metal deposition subsystems, and the VMC. In efforts to eliminate the separate VMC computer, required only to upload machine code via direct numerical control, the RS-232 communication protocol utilized by Fadal was reverse engineered and implemented via LABVIEW. The laser, cooling, and powder material delivery subsystems of the laser metal deposition process are equipped with external control ports, but were not utilized in previous system configurations. Subsequently, all subsystems and modules were directly connected to the control system hardware so external control could be utilized. Initializing communications among the LAMP subsystems became the foundation for the control system software. Off-line, the in-house layered manufacturing software only converted CAD models into layer-by-layer slices of machine code to create the tool path. With the central control system now in place, the in-house layered manufacturing software was changed to generate machine code, laser power, and powder flow commands, which together comprise a part program and are distributed via the control system software. Overall, manufacturing process integration has resulted in modularity, easy maintenance, and process improvement. Thus, increasing system productivity and capability.

5.3 Quantitative modeling and simulation

Quantitative modeling and simulation provides a theoretical foundation for explaining the phenomena observed through empirical research. Additionally, detailed modeling assists with developing a quantitative understanding of the relationship between independent process parameters and dependent process parameters. Understanding the relationships among parameters affords accurate control of physical dimension and material properties of the part. While separate modeling efforts were undertaken, outputs of one model feed into another. The following subsections summarize how quantitative modeling has been used to develop a theoretical understanding of the LAMP hybrid manufacturing process.

5.3.1 Melt pool modeling and simulation

Melt pool geometry and thermal behavior control are essential in obtaining consistent building performances, such as geometrical accuracy, microstructure, and residual stress. A 3D model was developed to predict the thermal behavior and geometry of the melt pool in the laser material interaction process (Han et al., 2005). The evolution of the melt pool and effects of the process parameters were investigated through modeling and simulations with stationary and moving laser beam cases.

When the intense laser beam irradiates on the substrate surface, the melt pool will appear beneath the laser beam and it moves along with the motion of the laser beam. In order to interpret the interaction mechanisms between laser beam and substrate the model considers the melt pool and adjacent region. The governing equations for the conservation of mass, momentum and energy can be expressed in following form:

$$\frac{\partial}{\partial t}(\rho) + \nabla \cdot (\rho \mathbf{V}) = 0 \tag{1}$$

$$\frac{\partial}{\partial t}(\rho \mathbf{V}) + \nabla \cdot (\rho \mathbf{V} \mathbf{V}) = \nabla \cdot (\mu_l \frac{\rho}{\rho_l} \nabla \mathbf{V}) - \nabla p - \frac{\mu_l}{K} \frac{\rho}{\rho_l}(\mathbf{V} - \mathbf{V}_s) + \rho g \tag{2}$$

$$\frac{\partial}{\partial t}(\rho h) + \nabla \cdot (\rho \mathbf{V} h) = \nabla \cdot (k \nabla T) - \nabla \cdot (\rho (h_l - h)(\mathbf{V} - \mathbf{V}_s)) \tag{3}$$

where ρ, **V**, p, μ, T, k, and h are density, velocity vector, pressure, molten fluid dynamic viscosity, temperature, conductivity, and enthalpy, respectively. K is the permeability of mushy zone, **V**s is moving velocity of substrate with respect to laser beam and subscripts of s and l represent solid and liquid phases. Since the solid and liquid phases may coexist in the same calculation cell at the mushy zone, mixed types of thermal physical properties are applied in the numerical implementation. The liquid/vapor interface is the most difficult boundary for numerical implementation in this model since many physical phenomena and interfacial forces are involved there. To solve those interfacial forces the level set method is employed to acquire the solution of the melt pool free surface (Han et al., 2005). To avoid numerical instability arising from the physical property jump at the liquid/vapor interface, the Heaviside function $H(\varphi)$ is introduced to define a transition region where the physical properties are mollified.

The energy balance between the input laser energy and heat loss induced by evaporation, convection and radiation determines surface temperature. Laser power, beam spot radius, distance from calculation cell to the beam center, and the absorptivity coefficient are used to calculate the laser heat influx. Heat loss at the liquid/vapor interface is computed in terms of convective heat loss, radiation heat loss and evaporation heat loss. The roles of the convection and surface deformation on the heat dissipation and melt pool geometry are revealed by dimensionless analysis. It was found that interfacial forces including thermo-capillary force, surface tension and recoil vapor pressure considerably affect the melt pool shape and fluid flow. Quantitative comparison of interfacial forces indicates that recoil vapor pressure is dominant under the melt pool center while thermo-capillary force and surface tension are more important at the periphery of the melt pool.

For verification, the intelligent vision system was utilized to acquire melt pool images in real time at different laser power levels and process speeds, and the melt pool geometries were measured by cross-sectioning the samples obtained at various process conditions (Han et al., 2005). Simulation predictions were compared to experimental results for both the stationary laser case and moving laser case at various process conditions. Model prediction results strongly correlate to experimental data. An example of melt pool shape comparison between simulation and experiment for the moving laser beam case is shown in Fig. 5.

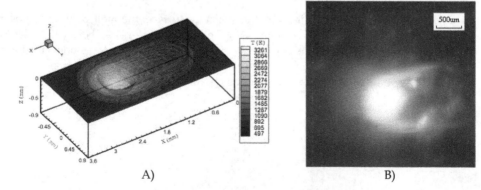

A)   B)

Fig. 5. Melt pool shape comparison, A) Simulation result of melt pool shape and surface temperature, B) Experimental result of melt pool shape (Adapted from Han et al., 2005)

5.3.2 Powder flow dynamics modeling and simulation

Analysis of metallic powder flow in the feeding system is of particular significance to researchers in order to optimize the LMD fabrication technique. Powder flow simulation holds a critical role in understanding flow phenomena. A stochastic Lagrangian model for simulating the dispersion behavior of metallic powder, or powder flow induced by non-spherical particle-wall interactions, is described (Pan & Liou, 2005). The numerical model also takes into consideration particle shape effects. In wall-bounded, gas-solid flows, the wall collision process plays an important role and is strongly affected by particle shape. Non-spherical effects are considered as the deviation from pure spheres shows induced particle dispersion, which has a great impact on the focusibility of the powder stream at the laser cladding head nozzle exit. The parameters involved in non-spherical collision are analyzed for their influencing factors as well as their interrelations.

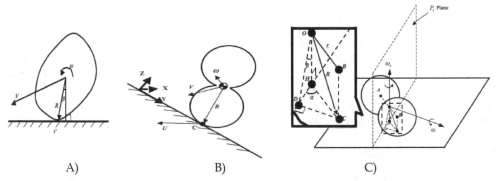

A)   B)   C)

Fig. 6. Particle Collision Diagrams, A) 2D non-spherical particle-wall collision model, B) Local coordinate for collision model, C) of 3-D non-spherical particle-wall collision model (Adapted from Pan & Liou, 2005)

The parameters involved in the 2-D non-spherical model include β and R, as shown in Fig. 6, Part A, where β indicates how much the contact point C deviates from the foot of a vertical from the gravity center of the particle and R shows the actual distance between the contact point and the gravity center. The collision coordinate system used to describe the 3D collision dynamics is defined in Fig. 6, Part B. The contact velocity is computed from:

$$U = V + \omega \times R \tag{4}$$

where V is the particle translational velocity vector, ω is the angular velocity vector, and R is the vector connecting particle mass center to contact point C. The change in the contact point velocity can be obtained by the following equation:

$$\Delta U = [\frac{1}{m} I - R^{\times} J^{-1} R^{\times}] \Delta P \tag{5}$$

where m is particle mass, I is the 3x3 identity matrix, and R^x is the canonical 3x3 skew-symmetric matrix corresponding to R, ΔP denotes the impulse delivered to the particle in the collision, and J^{-1} is the inverted inertia tensor in the local coordinate. As shown in Fig. 6,

Part C, a cluster that consists of two identical spheres with equal radius r represents the non-spherical powder particle of the 3D model. This representation leads to generalized modeling of satelliting metallic powder particles.

Wall roughness also effects powder dispersion behavior, therefore in this model the roughness effect was included by using the model and parameters proposed by (Sommerfeld & Huber, 1999). The instantaneous impact angle is assumed to be composed of the particle trajectory angle with respect to the plane wall and a random component sampled from a Gaussian distribution function. It was also assumed that each collision has 30% possibility to be non-spherical, which implies the stochastic model was applied in 30% of the total collisions during the feeding process simulation. Simulations using the spherical model (0% non-sphericity) were also conducted.

The non-spherical model successfully predicts the actual powder concentration profile along the radial and axial directions, whereas the spherical particle model underestimates the dispersion and results in a narrow spread of the stream along the radial direction. When compared to the experimental results, the 3D simulated powder stream is in strong agreement, which demonstrates validation of the model. The model also predicts the peak powder concentration or focal point of the power stream for specific cladding head nozzle geometry. It is essential to establish a well-focused powder stream at the exit of the nozzle and to know the ideal stand-off distance in order to increase powder catchment in the melt pool, achieve high material integrity, and reduce material waste.

5.3.3 Tool path modeling and simulation

Process planning, simulation, and tool path generation allows the designer to visualize and simulate part fabrication prior to manufacturing to ensure a successful process. Adaptive multi-axis slicing, collision detection, and adaptive tool path pattern generation for LMD as well as tool path generation for surface machining are the key advantages to the integrated process planning software developed for the LAMP hybrid system (Ren et al., 2010).

Basic planning steps involve determining the base face and extracting the skeleton of an input CAD model (Fig. 7, top left). The skeleton is found using the centroidal axis extraction algorithm (Fig. 7, top right). Based on the centroidal axis, the part is decomposed into sub-components and for each sub-component a different slicing direction is defined according to build direction. In order to build some of the components, not only translation but also rotation will be needed to finish building the whole part because different sub-components have different building directions (Fig. 7, bottom middle), and the laser nozzle direction is always along the z-axis. After the decomposition (Fig. 7, bottom left) results are obtained, the relationship among all the components is determined, and a building relationship graph is created.

From the slicing results and build directions, collision detection is determined. Collision detection is implemented by Boolean operation, which is an intersection operation, on a simulation (Ren et al., 2010). If the intersection result of the updated CAD model and the cladding head nozzle is not empty, then collision will happen in the real deposition process. The deformation of the CAD model following the building relationship graph includes two categories: positional deformation and dimensional deformation. Positional change means translation or rotation of the CAD model. The dimensions will change after every slicing layer is finished. For every updated model, collision needs to be checked before the next slicing layer is added. Following the collision detection algorithm, if a potential collision is detected the sequence of the slicing layers is reorganized (Ren et al., 2010). The output of the collision detection algorithm will be the final list of slicing layers, which comprise the actual building sequence when manufacturing the part.

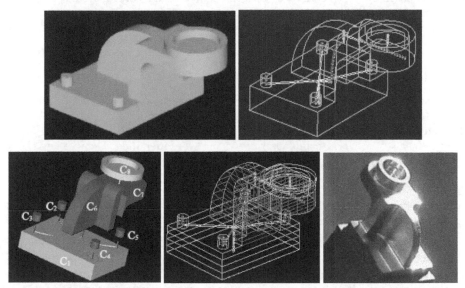

Fig. 7. Process Planning and Fabrication of 3D Part (Adapted from Liou et al., 2007; Ren et al. 2010)

The final piece of process planning is tool path generation. Common tool path patterns are the raster, contour-parallel offsetting, zig-zag, and interlaced. Each pattern has advantages and disadvantages. The adaptive deposition tool path algorithm considers each pattern when predicting the possibility of deposition voids. The goals of the algorithm are to adjust the tool path to remove deposition voids and increase time efficiency. Multiple tool path patterns may be used during fabrication and the algorithm may also prescribe alternating the appropriate tool path pattern when necessary.

Surface finish machining is a sequential step used after deposition to improve manufacturing quality after deposition is finished. The process planning software allows the designer to specify the machining parameters including the feed rate, spindle speed, and depth of cut before determining the number of machining cycles necessary. As with LMD, alignment will be also integrated for 3D geometries to achieve the accuracy without reloading the deposited part to be machined. Again, the tool path will be generated such that a collision-free machining tool path will be generated for the deposited part. A visibility map algorithm (Ruan & Liou, 2003) is applied to detect the collision between the tool and the deposited part.

The final process planning step is to generate the part program. This step is the bridge between the algorithmic results of process planning, quantitative modeling of process parameters, and the realistic operational procedures as well as parameters of the 5-axis manufacturing environment. It will build the map of the process planning results and the real operational parameters and then interpret the final planning tool path as the corresponding movements of the hybrid manufacturing system. The software will combine and refine those movements and translate them into machine executable code. Resulting in a text file composed of three columns of data to needed for the control system to command the laser, powder feeder, and motion system (Ren et al., 2010). The final set of operations is based on the building relationship graph, build directions that avoid collisions, tool path, and time required.

6. Hybrid manufacturing system integration

During the course of this research several integrated manufacturing system designs were analyzed to identify what characteristics comprise a successful hybrid system. Based on this background research, and the experiences of working with and refining the LAMP system, the key elements of a hybrid manufacturing system were identified. The five key elements represent an effective way to design a hybrid manufacturing system, as compared to a reconfigurable or mechatronic design, because the identified elements contain necessary subsystems, are easily modularized, and advocate the use of off-the-shelf hardware and software. Within an integrated system, each element acts as a separate subsystem affording a stable modular design (Gerelle & Stark, 1988).

A strategy for controlling the integrated LMD and machining processes, the 5-axis motion system, and the data corresponding streams provides the basis for fully automating the system. Considering scalability, our integration strategy emphasizes modularity of the integrated components but also modularity of the controlling software. Our control strategy allows data streams to be easily added or removed. Furthermore, our design allows an operator to optimize the control strategy for a particular geometry.

6.1 Physical Integration

Obstacles arise during the development of any manufacturing system; however, by identifying obstacles and solutions the industry as a whole can benefit. Outside of cost and yield, the obstacles of developing a hybrid manufacturing system discussed here cover a range of topics. Table 2 summarizes the obstacles associated with the physical integration of the LAMP hybrid system and provides documented solutions. The documented information in Table 2 does not address every possible integration obstacle, but is meant to be comprehensive from what is found in the literature and personal experience. Issues outside of integration, such as material properties can be found in (Nagel & Liou, 2010).

After central control, integration, and modularity were enforced in the LAMP hybrid system, manufacturing defects and time were significantly reduced, and safety was significantly increased. Material integrity was improved as the laser could be precisely commanded on/off or pulsed as needed during deposition. Furthermore, by integrating the laser power and powder flow commands into the process planning software and automating the distribution of commands, functionally graded parts were manufactured effortlessly.

Issue	Solution	Result	Reference
Adding the laser cladding head to a VMC	A platen with precisely tapped holes for the cladding head mounted to the Z-axis of the VMC	Laser cladding head is securely mounted and future equipment or fixtures can be added	
Protection of Equipment	Retract laser head or position it far enough away from the machining head	Protect laser nozzle	Kerschbaumer & Ernst, 2004
	Mount a displacement sensor on the Z-axis	When cladding head gets too close to X-Y axes the process halts	
Unknown communication protocol	Use reverse engineering to figure out communication protocol	Subsystems can be controlled from a central control system	Stroble et al., 2006
Quality control	Implement control charts, pareto charts, etc.	Manual quality control	Starr, 2004
	Sensor feedback utilized by closed-loop controllers	Automated quality control	Boddu et al., 2003; Doumanidis & Kwak, 2001; Hu & Kovacevic, 2003; Tang, 2007
Transition between additive and subtractive processes	Apply a translation matrix that repositions the X-Y axis for the desired process	Accurate positioning for machining or LMD	
Placement of sensors to monitor melt pool due to high heat of the LMD process	Mount the sensitive vision system in-line with the laser using a dichromatic mirror attachment for the cladding head, and custom hardware mounted to the platen holds the temperature probe at an acceptable viewing angle	Sensors are safe, and the LMD process is accessible	Boddu et al., 2003; Tang & Landers, 2010

Table 2. Physical Integration Issues and Solutions

6.2 Software Integration

Utilization of a central control system directly resulted in automation of the LAMP hybrid system and allowed unconventional possibilities to be explored. To achieve the central controller, a framework consisting of a multi-phase plan and implementation methodology was developed. The automation framework involves controlling the laser, powder feeder, and motion system, and utilizing sensor feedback, all through the NI PXI control subsystem. Open and closed-loop controllers were designed, along with compatibility and proper module communication checking. Moreover, compensation for undesired system dynamics, delays and noise were considered to ensure a reliable and accurate automated manufacturing process. The result of the automation framework is an automated deposition program (developed in LabVIEW) with a customized graphical user interface and data recording capabilities.

Figure 8 is a visual description of the LAMP hybrid system communications layout, including process planning that occurs outside the control system. Once process planning completes the part program, with laser power and powder mass flow rate commands in the form of voltages, the control system parses through the information to automatically fabricate the desired part. While commands are being sent to the physical devices, sensors are monitoring the process and sending feedback to the control system simultaneously, allowing parameters to change in real-time.

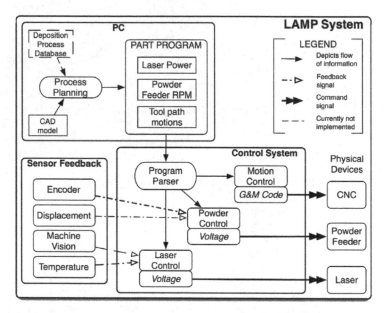

Fig. 8. LAMP Hybrid System Communication Schematic

Unique to the LAMP hybrid system is that the hardware and software are both modular. The automated deposition program that is executed by the control system has three different modes: dry-run, open-loop control, and closed-loop control. Fundamental code within the automated deposition program is shared amongst each of the modes, much like the control system is central to the LAMP hybrid system. Additional portions of code that control the laser, control the powder feeder, utilize feedback, or simply read in, display and record data from sensors are turned on or off by each mode. Code modularity prevents large amounts of the control system software from being rewritten when equipment is upgraded or subsystems are replaced.

During dry-run mode only machine code is distributed by the control system, allowing the user to monitor the VMC motions without wasting materials and energy. This mode is primarily utilized to check uncertain tool paths for instances when the laser should be shut off or when a tool path transition seems too risky. For instance, transitions from one geometry to another may rotate longer than desired at one point causing a mound to form and solidify, which destroys the overall part geometry and could collide with the laser cladding nozzle. Open-loop and closed-loop control modes are provided for fabricating parts and include system monitoring and data acquisition features. The modular software allows multiple closed-loop controllers optimized for a particular geometry to be added as research is completed, such as a feed forward controller (Tang et al., 2007) that regulates powder flow to the melt pool for circular, thin walled structures or thin walled structures with many arcs.

7. Conclusion

In an effort to shorten the time-to-market, decrease the manufacturing process chain and cut production costs, research has focused on the integration of multiple manufacturing processes into one machine; meaning less production space, time, and manpower needed. An integrated or hybrid system has all the same features and advantages of rapid prototyping systems, plus provides a new set of features and benefits. Moreover, hybrid manufacturing systems are increasingly being recognized as a means to produce parts in material combinations not otherwise possible and have the ability to fabricate complex internal geometries, which is beyond anything that can be accomplished with subtractive technologies alone. Internal geometries such as complex conformal cooling channels provide better product thermal performance, which additive fabrication processes create them with ease, giving the manufacturer a better product with little extra cost. As manufacturers and customers dream up more complex products, requiring more advanced equipment and software, hybrid systems will emerge. In short, integrating additive and subtractive technologies to create new manufacturing systems and processes is going to advance the manufacturing industry in today's competitive market.

Modeling and simulation, both qualitative and quantitative, were shown to be an integral part of hybrid system design and development as well as motivate areas of research that a pure empirical approach does not reveal. Although this research is focused on integrating additive and subtractive processes, the general principles can also be applicable to integrating other unit manufacturing processes (NRC, 1995). Integrated processes can combine multiple processes that fall within the same family, such as different material removal processes, or they can combine processes that are in different unit process families, such as a mass-change process and a microstructure-change process. The results can lead to significant processing breakthroughs for low-cost, high-quality production.

Future work includes applying integrated process and product analysis to various hybrid processes that integrate different manufacturing processes and applying the hybrid system concept to other types of configurations, such as those that include robots. Model-based simulation reveals various new opportunities for simultaneous improvement of part quality, energy and material efficiencies, and environmental cleanness. Thereby, accelerating the hybrid integration process. Other work includes applying an open architecture for the hybrid controller, as such an architecture avoids the difficulties of using proprietary technology and offers an efficient environment for operation and programming, ease of integrating various system configurations, and provides the ability to communicate more effectively with CAD/CAM systems and factory-wide information management systems.

8. Acknowledgment

This research was supported by the National Science Foundation Grant # IIP-0822739, the U.S. Air Force Research Laboratory, and the Missouri S&T Intelligent Systems Center. We would also like to thank all the Missouri S&T LAMP lab researchers that have contributed over the years to make this body of work possible, especially Zhiqiang Fan, Lijun Han, Heng Pan, Lan Ren, Jianzhong Ruan, and Todd Sparks. Their support is greatly appreciated.

9. References

Akula, S. & Karunakaran, K.P. (2006). Hybrid Adaptive Layer Manufacturing: An Intelligent Art of Direct Metal Rapid Tooling Process. *Robotics and Computer-Integrated Manufacturing*, Vol. 22, No. 2, pp. 113-123, 0736-5845

Boddu, M.R., Landers, R.G., Musti, S., Agarwal, S., Ruan, J. & Liou, F. W. (2003) System Integration and Real-Time Control Architecture of a Laser Aided Manufacturing Process. *Proceedings of SFF Symposium*, 1053-2153, Austin, TX, August, 2003

Choi, D.-S., Lee, S.H., Shin, B.S., Whang, K.H., Song, Y.A., Park, S.H. & Lee, H.S. (2001) Development of a Direct Metal Freeform Fabrication Technique Using Co2 Laser Welding and Milling Technology. *Journal of Materials Processing Technology*, Vol. 113, No. 1-3, (June 2001), pp. (273-279), 0924-0136.

Cooper, A.G. (1999). Fabrication of Ceramic Components Using Mold Shape Deposition Manufacturing. Doctor of Philosophy Thesis, Stanford, USA.

Dorf, R.C. & Kusiak, A. (1994). *Handbook of Design, Manufacturing and Automation*, J. Wiley and Sons, 0471552186, New York, N.Y.

Doumanidis, C. & Kwak, Y.-M. (2001) Geometry Modeling and Control by Infrared and Laser Sensing in Thermal Manufacturing with Material Deposition. *Journal of Manufacturing Science and Engineering*, Vol. 123, No. 1, pp. 45-52, 1087-1357

Duley, W.W. (2003). *Laser Processing and Analysis of Materials*, Plenum Press, 0306410672, New York

Fessler, J.R., Merz, R., Nickel, A.H. & Prinz, F.B. (1999) Laser Deposition of Metals for Shape Deposition Manufacturing. *Proceedings of SFF Symposium*, 1053-2153, Austin, TX, August, 1999

Gebhardt, A. (2003). *Rapid Prototyping*, Hanser Publications, 9781569902813, Munich

Gerelle, E.G.R. & Stark, J. (1988). *Integrated Manufacturing: Strategy, Planning, and Implementation*, McGraw-Hill, 0070232350, New York

Han, L., Liou, F. & Musti, S. (2005). Thermal Behavior and Geometry Model of Melt Pool in Laser Material Process. *Journal of Heat Transfer*, Vol. 127, No. 9, pp. 1005, 0022-1481

Hopkinson, N., Hague, R.J.M. & Dickens, P.M. (2006) *Rapid Manufacturing: An Industrial Revolution for the Digital Age*, John Wiley, 0470016132, Chichester, England

Hu, D. & Kovacevic, R. (2003). Sensing, Modeling and Control for Laser-Based Additive Manufacturing. *International Journal of Machine Tools & Manufacture*, Vol. 43, No. 1, (January 2003), pp. 51-60, 0890-6955

Hu, D., Mei, H. & Kovacevic, R. (2002). Improving Solid Freeform Fabrication by Laser-Based Additive Manufacturing. *Proceedings of the Institution of Mechanical Engineers, Part B: Journal of Engineering Manufacture*, Vol. 216, No. 9, pp. 1253-1264, 1253-1264

Hur, J., Lee, K., Hu, Z. & Kim, J. (2002). Hybrid Rapid Prototyping System Using Machining and Deposition. *Computer-Aided Design*, Vol. 34, No.10, (September 2002), pp. 741-754, 0010-4485

Jeng, J.-Y. & Lin, M.-C. (2001). Mold fabrication and modification using hybrid processes of selective laser cladding and milling. *Journal of Materials Processing Technology*, Vol. 110, No. 1, (March 2001), pp 98-103, 0924-0136

Kai, C.C., & Fai, L.K. (1997). *Rapid Prototyping: Principles & Applications in Manufacturing*, John Wiley, 9810245165, New York.

Kerschbaumer, M., & Ernst, G. (2004). Hybrid Manufacturing Process for Rapid High Performance Tooling Combining High Speed Milling and Laser Cladding. *Proc. 23rd International Congress on Applications of Lasers & Electro-Optics (ICALEO)*, 0912035773 , San Francisco, CA, October 2004

Klocke, F. (2002). *Rapid Manufacture of Metal Components*. Fraunhofer Institute for Production Technology, IPT

Kalpakjian, S., & Schmid, S.R. (2003). *Manufacturing Processes for Engineering Materials*. Pearson Education, Inc., 9780130453730, Upper Saddle River, N.J

Lin, J. & Steen, W.M. (1998). Design characteristics and development of a nozzle for coaxial laser cladding. *Journal of Laser Applications*, Vol. 10, No. 2, pp. 55-63, 1042-346X

Liou, F.W., & Kinsella, M. (2009). A Rapid Manufacturing Process for High Performance Precision Metal Parts. *Proceedings of SME Rapid 2009 Conference and Exhibition*, Paper No. TP09PUB18, Schaumburg, IL, May, 2009

Liou, F., Slattery, K., Kinsella, M., Newkirk, J., Chou, H.-N., Landers, R. (2007). Applications of a Hybrid Manufacturing Process for Fabrication and Repair of Metallic Structures, *Rapid Prototyping Journal*, Vol. 13, No. 4, pp. 236–244, 1355-2546

Liou, F.W., Choi, J., Landers, R.G., Janardhan, V., Balakrishnan, S.N., & Agarwal, S. (2001). Research and Development of a Hybrid Rapid Manufacturing process. *Proceedings of SFF Symposium*, 1053-2153, Austin, TX, August, 2001

Mazumder, J., Choi, J., Nagarathnam, K., Koch, J. & Hetzner, D. (1997). The Direct Metal Deposition of H13 Tool Steel for 3-D Components. *JOM*, Vol. 49, No. 5, pp.55-60 1047-4838

Mazumder, J., Schifferer, A., & Choi, J. (1999). Direct Materials Deposition: Designed Macro and Microstructure", Materials Research Innovations, Vol. 3, No. 3, (October 1999), pp.118-131, 1432-8917

NRC (National Research Council) (1995). *Unit Manufacturing Processes: Issues and Opportunities in Research*, The National Academies Press

Nagel, J.K.S. & Liou, F. (2010). Designing a Modular Rapid Manufacturing Process. *Journal of Manufacturing Science and Engineering*, Vol. 132, No. 6, (December 2010), pp. 061006, 1087-1357

Nagel, R., Hutcheson, R., Stone, R., & Mcadams, D., (2009). Process and Event Modeling for Conceptual Design. *Journal of Engineering Design*, Vol. 22, No. 3, (March 2011), pp.145-164, 0954-4828

Nowotny, S., Scharek, S., & Naumann, T. (2003). Integrated Machine Tool for Laser Beam Cladding and Freeforming. *Proc. Thermal Spray 2003: Advancing the Science & Applying the Technology*, 9780871707857, Orlando, FL, May, 2003

Pan, H. & Liou, F. (2005). Numerical simulation of metallic powder flow in a coaxial nozzle for the laser aided deposition process. *Journal of Materials Processing Technology*, Vol. 168, No. 2, pp. 230-244, 0924-0136

Ren, L., Sparks, T., Ruan, J. & Liou, F. (2010). Integrated Process Planning for a Multiaxis Hybrid Manufacturing System. *Journal of Manufacturing Science and Engineering*, Vol. 132, No. 2, pp. 021006, 1087-1357

Ren, L., Padathu, A.P., Ruan, J., Sparks, T., & Liou, F.W. (2008). Three Dimensional Die Repair Using a Hybrid Manufacturing System. *Proceedings of SFF Symposium*, 1053-2153, Austin, TX, August, 2008

Ruan, J., Eismas-Ard, K., & Liou, F. W., 2005, "Automatic Process Planning and Toolpath Generation of a Multiaxis Hybrid Manufacturing System," Journal of Manufacturing Processes, 7(1), pp. 57-68.

Ruan, J., & Liou, F.W. (2003) Automatic Toolpath Generation for Multi-axis Surface Machining in a Hybrid Manufacturing System. *Proc. of ASME IDETC/CIE*, 0791837009, Chicago, IL., September, 2003

Shunk, D.L. (1992). *Integrated Process Design and Development*, Business One Irwin, 9781556235566, Homewood, IL.

Sommerfeld, M., & Huber, N. (1999). Experimental analysis and modeling of particle-wall collisions. *International of multiphase flow*, Vol. 25, pp.1457-1489, 0301-9322

Song, Y.-A., & Park, S. (2006). Experimental Investigations into Rapid Prototyping of Composites by Novel Hybrid Deposition Process. *Journal of Materials Processing Technology*, Vol. 171, No. 1, (January 2006), pp. 35-40, 0924-0136

Starr, M. K., 2004, *Production and Operations Management*, Atomic Dog, 1592600921, Cincinnati, OH.

Stroble, J.K., Landers, R.G., & Liou, F.W. (2006). Automation of a Hybrid Manufacturing System through Tight Integration of Software and Sensor Feedback. *Proceedings of SFF Symposium*, 1053-2153, Austin, TX, August, 2006

Tang, L. & Landers, R.G. (2010). Melt Pool Temperature Control for Laser Metal Deposition Processes, Part I: Online Temperature Control," *Journal of Manufacturing Science and Engineering*, Vol. 132, No. 1, (February 2010), pp. 011010, 1087-1357

Tang, L., Ruan, J., Landers, R., & Liou, F. (2007). Variable Powder Flow Rate Control in Laser Metal Deposition Processes, *Proceedings of SFF Symposium*, 1053-2153, Austin, TX, August, 2007

Venuvinod, P.K., & Ma, W. (2004). *Rapid Prototyping Laser-Based and Other Technologies*, Kluwer Academic Publishers, ISBN: 978-1-402-07577-3, Boston

Wang, B. (1997). *Integrated Product, Process and Enterprise Design*, Chapman & Hall, 0412620200, New York.

Weerasinghe V.W. & Steen, W.M. (1983). Laser Cladding with Pneumatic Powder Delivery. *Proceedings of Laser Materials Processing*, Los Angeles, CA, January 1983

Westkämper, E. (2007). Digital Manufacturing In The Global Era, In: *Digital Enterprise Technology*, Pedro Cunha and Paul Maropoulos, pp. 3-14, Springer, 978-0-387-49864-5, New York

Xue, L. (2006). Laser Consolidation – a One-Step Manufacturing Process for Making Net-Shaped Functional Aerospace Components. *Proc. SAE International Aerospace Manufacturing and Automated Fastening Conference & Exhibition*, Toulouse, France, September, 2006

Xue, L. (2008). *Laser Consolidation--a Rapid Manufacturing Process for Making Net-Shape Functional Components*, Industrial Materials Institute, National Research Council of Canada, London, Ont.

Yang, D.Y., Kim, H.C., Lee, S.H., Ahn, D.G. & Park, S.K. (2005). Rapid Fabrication of Large-Sized Solid Shape Using Variable Lamination Manufacturing and Multi-Functional Hotwire Cutting System. *Proceedings of SFF Symposium*, 1053-2153, Austin, TX, August, 2005

3axis, Direct Metal Laser Sintering (Dmls-Eos), Retrieved on Jan. 14, 2010, http://www.3axis.us/direct_metal_laser_slintering_dmls.asp

Permissions

The contributors of this book come from diverse backgrounds, making this book a truly international effort. This book will bring forth new frontiers with its revolutionizing research information and detailed analysis of the nascent developments around the world.

We would like to thank Engr. Dr. Faieza Abdul Aziz, Dr. Azmah Hanim Mohamed Ariff, Dr. Azfanizam Ahmad, Ir.N.Jayaseelan and Dr. Qaisar Ahsan , for lending their expertise to make the book truly unique. They have played a crucial role in the development of this book. Without their invaluable contribution this book wouldn't have been possible. They have made vital efforts to compile up to date information on the varied aspects of this subject to make this book a valuable addition to the collection of many professionals and students.

This book was conceptualized with the vision of imparting up-to-date information and advanced data in this field. To ensure the same, a matchless editorial board was set up. Every individual on the board went through rigorous rounds of assessment to prove their worth. After which they invested a large part of their time researching and compiling the most relevant data for our readers. Conferences and sessions were held from time to time between the editorial board and the contributing authors to present the data in the most comprehensible form. The editorial team has worked tirelessly to provide valuable and valid information to help people across the globe.

Every chapter published in this book has been scrutinized by our experts. Their significance has been extensively debated. The topics covered herein carry significant findings which will fuel the growth of the discipline. They may even be implemented as practical applications or may be referred to as a beginning point for another development. Chapters in this book were first published by InTech; hereby published with permission under the Creative Commons Attribution License or equivalent.

The editorial board has been involved in producing this book since its inception. They have spent rigorous hours researching and exploring the diverse topics which have resulted in the successful publishing of this book. They have passed on their knowledge of decades through this book. To expedite this challenging task, the publisher supported the team at every step. A small team of assistant editors was also appointed to further simplify the editing procedure and attain best results for the readers.

Our editorial team has been hand-picked from every corner of the world. Their multi-ethnicity adds dynamic inputs to the discussions which result in innovative outcomes. These outcomes are then further discussed with the researchers and contributors who give their valuable feedback and opinion regarding the same. The feedback is then collaborated with the researches and they are edited in a comprehensive manner to aid the understanding of the subject.

Apart from the editorial board, the designing team has also invested a significant amount of their time in understanding the subject and creating the most relevant covers. They scrutinized every image to scout for the most suitable representation of the subject and create an appropriate cover for the book.

The publishing team has been involved in this book since its early stages. They were actively engaged in every process, be it collecting the data, connecting with the contributors or procuring relevant information. The team has been an ardent support to the editorial, designing and production team. Their endless efforts to recruit the best for this project, has resulted in the accomplishment of this book. They are a veteran in the field of academics and their pool of knowledge is as vast as their experience in printing. Their expertise and guidance has proved useful at every step. Their uncompromising quality standards have made this book an exceptional effort. Their encouragement from time to time has been an inspiration for everyone.

The publisher and the editorial board hope that this book will prove to be a valuable piece of knowledge for researchers, students, practitioners and scholars across the globe.

List of Contributors

Barthélemy H. Ateme-Nguema
Université du Québec en Abitibi-Témiscamingue, Canada

Félicia Etsinda-Mpiga, Thiên-My Dao and Victor Songmene
École de Technologie Supérieure, Montreal, Canada

Faieza Abdul Aziz, Izham Hazizi Ahmad, Norzima Zulkifli and Rosnah Mohd. Yusuff
Universiti Putra Malaysia, Malaysia

Aslı Aksoy and Nursel Öztürk
Uludag University Department of Industrial Engineering, Bursa, Turkey

Jorge Cortés, Ignacio Varela-Jiménez and Miguel Bueno-Vives
Tecnológico de Monterrey, Campus Monterrey, México

Minna Lanz, Eeva Jarvenpaa, Fernando Garcia, Pasi Luostarinen and Reijo Tuokko
Tampere University of Technology, Finland

Namhun Kim
Ulsan National Institute of Science and Technology, Korea

Richard A. Wysk
North Carolina State University, USA

Diogo L. L. da Cruz
Department of Electrical Engineering, Santa Catarina State University, Brazil
Pollux Automation, Brazil

André B. Leal
Department of Electrical Engineering, Santa Catarina State University, Brazil

Marcelo da S. Hounsell
Department of Computer Science, Santa Catarina State University, Brazil

Michael Mutingi
National University of Singapore, Electrical & Computer Engineering, Singapore

Godfrey C. Onwubolu
School of Applied Technology, HITAL, Toronto, Canada

Tomasz Mączka and Tomasz Żabiński
Rzeszów University of Technology, Poland

Jacquelyn K. S. Nagel
James Madison University, USA

Frank W. Liou
Missouri University of Science and Technology, USA